Reaching and Teaching Neurodivergent Learners in STEM

Providing salient stories and practical strategies, this book empowers educators to embrace the unique talents of neurodivergent learners in science, technology, engineering, and mathematics (STEM). An exploration of the exciting opportunities neurodiversity presents to build an innovative workforce is grounded in a large body of research from psychology, neuroscience, and education.

Author Jodi Asbell-Clarke presents individual examples of neurodivergent journeys in STEM to establish evidence-based connections between neurodiversity and the types of innovative problem-solving skills needed in today's workforce. The featured stories come directly from the author's many years in inclusive classrooms with STEM teachers, along with interviews from many neurodivergent professionals in STEM. Teachers will learn how to embrace the unique brilliance and potential of the neurodivergent learners in their classroom, working against historic marginalization and deficit-based perspectives of neurodiversity within the education system.

Featuring illustrations of classroom-designed tools and materials alongside basic strategies to support executive function and emotion in learning, this book will help you nurture the talents of your neurodivergent learners and recognize their unique potential within STEM. Ideal for K–12 classroom teachers, special educators, learning specialists, psychologists, and school administrators.

Jodi Asbell-Clarke is a Senior Leader at the Center for Science Teaching and Learning at TERC, a nonprofit research and development firm focusing on innovative STEM education. She holds a PhD in education and was formerly a computer programmer and a physics and astrophysics teacher at University High School at the University of Illinois, USA.

"Imagine a world where the unique STEM-related talents of neurodivergent individuals are cultivated to grow a dynamic, innovative workforce. In her groundbreaking work, Jodi Asbell-Clarke presents an empowering roadmap for educators to embrace and nurture these talents, instead of perpetuating shame and pity".
—**Professor Sara Seager**, *Massachusetts Institute of Technology.*

"*Neurodivergent Learners in STEM*" is an empowering guide for educators to create inclusive learning environments, expertly navigating the intersection of STEM education and neurodiversity and providing practical strategies to unlock the potential of neurodivergent learners."
—**Matthew Farber**, author of *Gaming SEL: Games as Transformational to Social and Emotional Learning*

"*Reaching and Teaching Neurodivergent Learners in STEM* is an important book for all educators. It encourages us to look at all students as having huge potential. Don't leave behind students who think differently. They may be the leaders of their generation."
—**Nadine Bonda**, PhD, former Superintendent

Reaching and Teaching Neurodivergent Learners in STEM

Strategies for Embracing Uniquely Talented Problem Solvers

Jodi Asbell-Clarke

NEW YORK AND LONDON

Designed cover image: © Getty Images

First published 2024
by Routledge
605 Third Avenue, New York, NY 10158

and by Routledge
4 Park Square, Milton Park, Abingdon, Oxon, OX14 4RN

Routledge is an imprint of the Taylor & Francis Group, an informa business

© 2024 TERC

The right of Jodi Asbell-Clarke to be identified as author of this work has been asserted in accordance with sections 77 and 78 of the Copyright, Designs and Patents Act 1988.

All rights reserved. No part of this book may be reprinted or reproduced or utilised in any form or by any electronic, mechanical, or other means, now known or hereafter invented, including photocopying and recording, or in any information storage or retrieval system, without permission in writing from the publishers.

Trademark notice: Product or corporate names may be trademarks or registered trademarks, and are used only for identification and explanation without intent to infringe.

ISBN: 978-1-032-56246-9 (hbk)
ISBN: 978-1-032-56247-6 (pbk)
ISBN: 978-1-003-43461-0 (ebk)

DOI: 10.4324/9781003434610

Typeset in Palatino
by Apex CoVantage, LLC

For Z, whose story sparked the idea for this book, and for his mother, who has been my steadfast support and inspiration all along the way.

And, of course, for my father whose smile and wisdom beamed down on me as I wrote every chapter.

Contents

Acknowledgements..................................xii

1 Introduction..1
   My Reasons for Writing This Book.....................1
   A Bit About Me 3
   A Bit About the Book................................5
      Hillside Junior High School.........................6
      Drawing From Research..............................9
      Layout of the Book 9
   Bibliography for Chapter 1 11

2 Neurodiversity—The Words We Use 12
   Caleb—A Missed Opportunity 12
   Why Words Matter 14
      Categories of Neurodiversity 15
      Complications With Categorization..................16
      Asset-based Perspective of Neurodiversity 17
   Maria and Paul—Coming to Terms 20
   What Is Normal?....................................22
      The Normalization of Education 23
      What Is Neurotypical?..............................24
      Example Number One 25
      Example Number Two 25
      Example Number Three..............................26
   Bibliography for Chapter 2 27

3 Neurodiversity—The Competitive Advantage..........29
   Dennis—A Journey From Dishwasher
      to Programmer.....................................29
   Neurodiversity in the STEM Workforce...............32
   Sara—A Journey to Genius 35
   Bibliography for Chapter 3 38

4 Learning in the Brain ... 39
Chase—A Sentry at the Door ... 39
Brain Structure and Function ... 41
    Neuroimaging ... 42
    Structure of the Brain ... 42
    Networks in the Brain ... 43
    The Limbic System ... 44
    Neuroplasticity ... 45
Executive Function ... 46
    Executive Function Inside the Brain ... 46
    Executive Function in the Classroom ... 47
Kiera and Kelly—Two Different Problem Solvers ... 49
Bibliography for Chapter 4 ... 51

5 Learning in the Classroom ... 53
Uni High School—Learning About Learning ... 53
Constructing Knowledge ... 55
    Roots of Constructivism ... 55
    Social Constructivism ... 55
Learner-centered Learning ... 58
Bibliography for Chapter 5 ... 60

6 The Role of Emotion in Learning ... 61
Ridge Road Elementary—A Grade 6 Project in the 1960s ... 61
Emotional Engagement in Learning ... 63
    Motivation ... 64
    Flow ... 64
Becoming a STEM Problem Solver ... 65
STEM Identity and Learner Agency ... 66
    Learner Autonomy ... 67
Anita—Embracing Rabbit Holes ... 69
Bibliography for Chapter 6 ... 71

7 Extraordinary Learners ... 73
Addison and Krystal—Bursting With Excitement ... 73
Twice-exceptional Learners ... 77
    Overexcitabilities ... 77

Creativity 79
Systems and Patterns 81
Productive Persistence 86
Ms Bradbury's Grade 7 Class–Drumming
    Through Transitions 87
Bibliography for Chapter 7 90

8 Neurodiversity and Collaboration 92
Jay—Just a Matter of Time 92
Communication 93
Empathy 96
    Theory of Mind and the Double-empathy Problem......97
Carleton—A Missed Opportunity............................98
Bibliography for Chapter 8 101

9 Inclusive STEM Teaching..............................102
A Classroom Observation—This Hour Has
    22 Minutes 102
Neurodiversity in Schools 107
Differentiation and Universal Design....................110
    CAST and UDL Guidelines 111
    Flexibility and Adaptability.......................112
    Differentiated Assessments 113
Jen—Getting to Completion...............................114
Bibliography for Chapter 9 115

10 Strategies for Inclusion—Project-based Learning......117
What Is Project-based Learning?.........................117
Project-based Learning at Hillside Junior High 119
    Gathering Resources and Equipment 120
    Setting Milestones................................121
    Finding Purpose 122
Eli—Giving New Life to an Old Computer.................123
Supporting Executive Function in
    Project-based Learning.............................125
    Tapping Into Students' Interests..................127
    School Design Project.............................127
    Supporting Planning and Organization 129

Supporting Transitions..............................130
　　　Fitting Into the Crowded School Day 132
　　Inclusive Assessment of Project-based Learning 133
　　　Measuring Progress 133
　　　Assessment in the Moment 136
　　Bibliography for Chapter 10 137

11 Strategies for Inclusion—Game-based Learning.......138
　　Let Me Know When She Stops Talking 138
　　What Is Game-based Learning? 139
　　　What *Portal* Taught Me..........................140
　　　Learning in Commercial Games 141
　　　Implicit Game-based Learning 141
　　Martian Boneyards..................................142
　　　Measuring Scientific Inquiry in *Martian Boneyards* 144
　　Leveling Up 146
　　　Building Game-based Learning Assessments........147
　　　Adaptive Games 149
　　　Bridging Implicit Game-based Learning
　　　　to Explicit Classroom Learning..................150
　　Bibliography for Chapter 11..........................151

**12 Strategies for Inclusion—Computational
　　Thinking...153**
　　Renee—Creating a World of Cats 153
　　Computational Thinking 155
　　Zoombinis...157
　　　Zoombinis Puzzle 1: Allergic Cliffs 158
　　　Zoombinis Puzzle 2: Pizza Pass 159
　　Miaka—Designing Executive Function Supports........161
　　INFACT...163
　　Bibliography for Chapter 12 165

13 Neurodiversity in STEM—The Actions We Take 167
　　Reflections..167
　　Strategies for Employers...........................168
　　Strategies for Teachers 169

Strategies for Designers, Administrators,
    and Policy Makers.................................172
    Inclusive Schools..................................173
    Building Community 174
What Can Be...177
Bibliography for Chapter 13 178

Acknowledgements

In spring 2022, I was fortunate to receive a sabbatical from my institution, TERC, to gather my thoughts and write this book. For nearly 60 years, TERC has been a stronghold in innovative research and development in STEM education, always keeping an eye towards social equity and reaching those who are often missed. I am proud to have spent most of my career at this fine organization. I hope my writing reflects on the integrity and thoughtfulness with which TERC conducts their mission of innovation and social justice in STEM education.

Throughout my career, I have received generous research support from the US National Science Foundation and the US Department of Education. I have had the privilege of working alongside designers, researchers, and educators who are some of the most talented (and the most fun) people I know. Foremost in my mind is my team at EdGE at TERC—Teon Edwards, Jamie Larsen, Elizabeth Rowe, Erin Bardar, Kelly Paulson, Tara Robillard, Ibrahim Dahlstrom-Hakki, Renee Pawlowski, and Zac Alstad. I could never, and would never, have done this research without them. And I am also indebted to Harvey Yazijian, my longstanding editor at TERC, whose never-ending talent and patience always make my ideas clearer while also making sure the hyphens go in the right places. Thank you all.

The Halifax Regional Council of Education and the students and staff of "Hillside Junior High School" opened their doors to my participation and research in their project-based learning endeavors for eight years. This opportunity is beyond what most educational researchers ever dream about. My work at Hillside has changed me in so many ways, and I am truly grateful.

Lastly, to my family and friends who have inspired me with your many unique talents, your wonderful differences, and your bountiful love. It has been particularly fun to share this journey with my husband David—who has been writing his own book at the same time—although his primer on astrophysical magnetohydrodynamics has a *slightly* different tone. And many thanks to my children, who are no longer children, and to my friends—you know who you are—who have kept me laughing, learning, and appreciating my good fortunes throughout it all.

1

Introduction

My Reasons for Writing This Book

Our world is facing immense global problems in climate, disease, energy, food, and politics—huge, gnarly problems. We will need our most innovative STEM problem solvers to get us out of this mess. We need systematic thinkers who pay attention to every detail, and we need creative thinkers who look at problems from new perspectives. We need innovative problem solvers who make new connections between old ideas and look at unintended as well as intended consequences. We need non-conformists who speak up for their solutions no matter what others think. And we need problem solvers who imagine beyond what others have already envisioned and who persist until they succeed.

These problem solvers are out there. They are in our classrooms waiting to learn, waiting to be educated. But sometimes we pay too much attention to trying to fix these learners rather than nurturing their problem-solving talents. This book is a call to change the way we think of some of our most innovative thinkers. It is a suggestion for how we can include learners who think differently—neurodivergent learners—to unleash their hidden talents in STEM problem solving.

Neurodiversity is a relatively new term, having been coined by sociologist Judy Singer in the late 1990s to describe the differences in how people think and learn. While educators and psychologists often talk about neurological differences such as autism, dyslexia, and ADHD as deficit-based learning disabilities or disorders, a neurodiversity perspective eschews these labels and emphasizes that each individual brain is unique with its own strengths and weaknesses. Neurodiversity describes a natural variation in the structure and functioning of human brains and considers the individual "fingerprints" of each person's brain, where one brain is no more correct or typical than the other. In 1998, Harvey Blume wrote in *The Atlantic*, "Neurodiversity may be every bit as crucial for the human race as biodiversity is for life in general. Who can say what form of wiring will prove best at any given moment?"

Top global STEM companies, such as Microsoft, SAP, and EY (formerly known as Ernst & Young), see neurodiversity as a competitive advantage. They seek the creativity, systematic and detailed thinking, and persistence that neurodivergent employees bring to their organizations. Yet while these and other STEM companies are looking to include more neurodivergent employees to increase innovation, our educational systems still often regard neurodivergent learners as broken and needing to be fixed.

There are many dedicated and talented people working in the education system to help struggling learners—general education teachers, special educators, learning specialists, paraprofessionals, speech therapists, and occupational therapists, just to name a few—and I admire them greatly. But many of these learners don't need to be fixed. Many don't want to be fixed. Many have talents surpassing their peers. They just don't do well at the game of school. School is built around a rigid model of learner that isn't conducive to helping some of our brightest and most innovative learners thrive.

In a 2006 TED talk (to date, the most viewed TED talk of all time), education scholar Sir Ken Robinson said: "Our education system has mined our minds in the way that we strip-mine the earth: for a particular commodity. And for the future, it won't serve us." Robinson described our schools as products of an industrial age where we sort our children by manufacture date and educate them with an assembly-line mentality. While some school systems have slowly moved away from that bleak factory vision—Robinson's call for school reform that puts creativity and flexibility at the forefront of education remains largely unanswered.

Many neurodivergent learners have great strengths that they can contribute to our workplaces, schools, and future society, but enabling this inclusion calls for a rethink in STEM education. Neurodiversity has much to offer in many other fields as well, but I stick primarily to STEM because I am a STEM educator and there is ample research to fill a book just in how to embrace neurodiversity in STEM teaching and learning, my area of expertise.

This book is *not* trying to say that every neurodivergent learner will or should become an expert computer programmer or research scientist if given some magical recipe for the "right" type of education. Nor is it saying that all neurodivergent learners have all these talents. Neurodiversity means just that—a diverse range of ways people's brains work. The nature and the severity of differences in brain function can manifest itself in a broad range of behavior, emotion, and cognition. There is no one solution for everyone. I hope, however, that this book gives everyone—no matter where they and their loved ones are along the many different dimensions of neurodiversity—some inspiration to go out and solve problems. Little problems. Big problems. They are all worth trying to solve. And we all have something to contribute.

A Bit About Me

I am a learning scientist, which means I study how people learn. I have spent the last three decades working with my team at TERC[1] to study STEM teaching and learning at the elementary and secondary school levels (K–12). Our research typically takes place in classrooms with teachers and students. Sometimes we study learning in museums and in out-of-school programs, and sometimes with parents and their children in their homes. The core of my research has always been about how to engage more learners in high-quality STEM learning.

FIGURE 1.1 Jodi Asbell-Clarke

I studied math and computer science in college and was fortunate enough to have a co-op job with IBM, so upon graduation I landed a plum position in IBM's federal systems division in Clear Lake, Texas. At 21 years old, I was a verification analyst on the onboard flight software for NASA's space shuttle program. I had gone to Houston thinking I might eventually become an astronaut. After a few years of working on the shuttle software, and when the Challenger explosion put the program on hiatus, I entered a Ph.D. program in astrophysics at the University of New Mexico. Along the way, I discovered that not only was my poor eyesight and lack of extraordinary physical discipline ill-suited for the astronaut program, I also wasn't really cut out for the astrophysics research track either. I was more curious about how people learned science than about practicing the science myself.

My professors encouraged me to go into teaching and I spent three years teaching physics and astrophysics at one of the most extraordinary schools in the country, University High School in Urbana, Illinois. During that time, I also started a family and my husband had landed a postdoctoral position at Harvard, so we eventually moved to Cambridge, MA. It was there that I found TERC, where I have worked as a learning science researcher ever since.

Early in my days at TERC, my husband got a faculty position at Saint Mary's University in Halifax, Nova Scotia, so I started telecommuting internationally to TERC. This was back in the 1990s, when we didn't have Zoom or any kind of video conferencing. The World Wide Web was still brand new, and some of us were still using Gopher and Fetch to send files back and forth. I was a remote worker long before it was a *thing*. But I got to keep doing my research and I got to grow a team of wonderful educators and researchers to do it with me. TERC is a fabulously flexible and innovative organization. I feel so lucky to work there. As they say up here in Nova Scotia, I feel like I "landed with my arse in the butter firkin."

Because of my background in astronomy, I am often asked to visit local K–12 classrooms. In 2014, I got such an invitation from a science teacher at a small school in Halifax that I call Hillside Junior High School, or just Hillside. That teacher is Ms Sadie Bradbury. The receipt of that email was a pivotal moment in my career and began the journey of this book.

Ms Bradbury responded to my email with a follow-up message saying: "I've just returned from NASA's Space Camp. I brought back a space suit! Who's going to wear it? You or me?" I immediately knew this was going to be fun. Ms Bradbury and I began our classroom journey together that day in 2014 and continued until the day she retired in 2022. During those eight years we taught side by side in her science and technology classes, experimenting with new teaching, learning, and assessment ideas to meet the diverse range of her students. We still get together regularly and reflect, often with insightful observations and a round of hearty laughter.

A Bit About the Book

This book has a selection of stories presented throughout the chapters. Many of these stories come from my experiences in Ms Bradbury's classroom at Hillside Junior High School.

FIGURE 1.2 Ms Sadie Bradbury engaging her students with science in 2014
Source: Photograph by the author

Others come from my personal experience and research in other classrooms.

Hillside Junior High School

Hillside is a small junior high school (about 135 students in grades 7–9) in a part of Halifax known for older apartment buildings, modest homes, and a large public housing sector. The neighborhood has deep roots, with many families living there over multiple generations. There is also a history of upheaval when, in the 1960s, the city uprooted historic Black communities and displaced them to the public housing units built within the previously predominantly white, working-class neighborhood of Hillside. This led to racial and social tensions, as well as the development of interesting close-knit relationships in the community. The neighborhoods around Hillside were also among the few affordable places near the city core for newly landed immigrants, often refugees (I use the past tense here because in the past couple of years a housing boom in the city has rapidly changed the affordability of the Hillside neighborhood). With such an interesting and rich history, I figured that my invitation to visit Hillside back in 2014 would offer a valuable learning experience, and I was not disappointed.

When Ms Bradbury moved to Halifax in 2008, Hillside was having a hard time attracting and keeping staff. Frequent police visits and fights gave the school what some considered a bad reputation. In addition, because the enrollment of the school was low, the salary for administration was reduced accordingly, so it wasn't thought of as a plum job in the district. But Hillside had the kids Ms Bradbury wanted to work with. When she arrived in Halifax, she walked into the principal's office and said, "You need me," and she was right. By the time I met her at Hillside six years later, Ms Bradbury was a teacher leader and was also the school's coordinator for integrated learning. She supported the design and piloting of the Nova Scotia Department of Education's new project-based learning curriculum in science and technology education. She was a key figure in the school.

Throughout this book, I tell stories and draw conclusions from my work with Ms Bradbury in her classroom. These stories are influenced by our own lived experiences. Ms Bradbury was raised in the UK and in Tanzania, and she taught school in England and Ontario, Canada before coming to Halifax. I have lived and worked in schools in the US, Canada, and Grenoble, France (where we lived for a year when my children were in grades 3 and 6). In addition, my team collaborated with game-based learning researchers in Finland, where we travelled to observe their fabulous education system. This global experience has given me a unique perspective on what is, and what can be, in schools.

I am a white woman who does not self-identify as neurodivergent. The same is true for Ms Bradbury. I have lived and worked closely with many neurodivergent colleagues, friends, and family members. My team at TERC includes neurodivergent researchers. For my research for this book, I have met with hundreds of people who identify as neurodivergent. There are no broad-brush statements to make about the people I spoke with. They are all different. In fact, that's the point.

Ms Bradbury's students came to school each morning with a diverse set of physical, cognitive, social, and emotional situations. Some of them had highly involved and/or highly educated parents. Others were in "the system" of foster care and/or shelters. Some had parents in jail. More than a few had witnessed chronic and/or acute violence near or in their homes. Some had had just arrived in a new country, sometimes escaping a traumatizing war, and having to learn a new language and customs while their families tried to make ends meet. Some of the students had diagnoses of autism, ADD, dyslexia, or a combination. Some exhibited similar indicators of neurodivergence but had no diagnosis at all. Ms Bradbury could not control many of these factors, but they were strong influences on the teaching and learning that could happen in her classroom.

What *was* within Ms Bradbury's control, for the most part, was the relationship she developed with each child who walked through her door. In many cases, this started just by listening. One September morning, a boy arrived in Ms Bradbury's grade 7 class, having just moved up from the local elementary school. He introduced himself to her by saying: "You should know, I am *not* my sister." His sister had been a top performing student in Ms Bradbury's classes in the previous years before moving on to high school. Without a moment's pause, Ms Bradbury replied in her exuberant British accent: "You are right, sir! You are *not* your sister. And you know what? I am really glad about that. I have already gotten to know your sister, but I don't know anything about you yet. And I am so excited that I get to learn about *you* this year!"

Ms Bradbury and I both like to think there were many children who benefited from our partnership in the classroom, yet we also both know that there were still many who slipped through the cracks. When Ms Bradbury retired, she and I found ourselves reflecting upon the methodologies, the innovations, and the strategies we picked up along the way, as well as all the ones we discarded. We thought about the children we missed—those students who came to school, eager to learn but not quite ready for the experience school had to offer them. Or perhaps school wasn't ready for the experience those children could have offered us.

Drawing From Research
I feel extremely fortunate to have had such a unique opportunity to learn from such a passionate and brilliant teacher, and the Hillside community. I was authorized with research approval from the Halifax Regional School Board to observe and interview Hillside teachers and their students from whom I'd garnered parental or guardian consent. Most of the student quotes and composite stories in this book arise from that research. I also draw from a decade of research with my team at EdGE. In addition, in preparation for this book, I conducted nearly 100 interviews with neurodivergent learners and STEM professionals and their families, friends, and employers. I met with coordinators of hiring programs for neurodiversity in four STEM companies, 27 employees in neurodiverse programs at STEM companies or universities, four employers engaged in neurodiverse hiring programs at STEM companies, nine parents of neurodivergent learners, and 47 individuals (ages 13–66) who have been diagnosed or identify themselves as neurodivergent.

In telling their stories, I have changed many of the names and, in some cases, synthesized details from several individuals into composite characters to retain confidentiality. The direct quotes are all real, but some of the details from different stories are blended into one to protect individuals' privacy. I use pseudonyms for many of the individuals in the stories and for Hillside Junior High School to protect the privacy of the students who were minors at the time as well as adults who asked to remain anonymous and/or could not provide consent to be identified.

Layout of the Book
The book is laid out in 13 chapters to explore how to nurture the unique STEM problem-solving talents of neurodivergent learners.

The book discusses reasons for considering neurodiversity in STEM, exploring the psychology and neuroscience of how people learn and the extraordinary talents associated with neurodiversity. The later chapters of the book introduce strategies for embracing neurodiversity in the teaching and learning of STEM problem solving.

Following this introductory Chapter 1, Chapter 2 introduces the terminology used in the study of neurodiversity and neurotypicality, pointing out the limitations, complexities, and even potential dangers of some pathologizing language used in current medical and educational systems.

Chapter 3 presents an asset-based perspective to neurodiversity and describes how some STEM companies see neurodiverse talent as a competitive advantage.

Chapter 4 is an overview of the structure and function of the human brain. The brain's networks used to process information and respond to our environment are explored, with a focus on executive function and its implications on learning.

Chapter 5 provides a grounding from psychology and learning sciences literature on how people learn.

Chapter 6 digs into the role of emotion in learning, exploring motivation, engagement, and flow, as well as identity, agency, and autonomy.

Chapter 7 explores some of the extraordinary talents associated with twice exceptional learners, such as detailed pattern recognition, systematic thinking, creativity, and persistence and how these assets align with STEM problem solving.

Chapter 8 discusses some of the complications in areas of communication and empathy in a neurodiverse learning environment.

Chapter 9 begins the discussion of neurodiversity in the classroom, giving a brief history along with a glimpse into a middle school neurodiverse inclusive classroom. Strategies such as differentiation and universal design for learning are introduced.

Chapter 10 describes how project-based learning was used to include neurodivergent learners at Hillside Junior High School. Examples of instructional and assessment materials are provided.

Chapter 11 discusses how my research team at TERC has used game-based learning to reach marginalized learners.

Chapter 12 highlights the potential for computational thinking to reveal and support STEM problem-solving practices with neurodivergent learners.

Chapter 13 includes recommendations for educators, administrators, parents, and policy makers, as well as neurodivergent learners themselves, on how to think differently about neurodiversity and STEM education so that we can help the next generation of innovative problem solvers thrive.

I have tried to structure the book, for the most part, as a conversation with you, the reader. Anyone who knows me knows that my conversations can be somewhat one-sided, and then when I catch myself, I interrupt my monologues with a question: *So, what are you thinking?* It is the phrase that rattles off my tongue when I realize I've been talking too long and I want to check in. In that vein, I have sprinkled *So, what are you thinking?* questions throughout the book so you can check in with yourself (and others) to reflect on what you just read. Consider it your part of the conversation. Enjoy.

Note

1   TERC is a non-profit in Cambridge, Massachusetts focusing on innovation and social equity in STEM education.

Bibliography for Chapter 1

Blume, H. (1998). Neurodiversity: On the neurological underpinnings of geekdom. *The Atlantic.* https://www.theatlantic.com/magazine/archive/1998/09/neurodiversity/305909/

Robinson, K. (2006). *Do schools kill creativity* [Video]. Ted Conferences. https://www.ted.com/talks/sir_ken_robinson_do_schools_kill_creativity/no-comments

Singer, J. (1998). *Odd people in: The birth of community amongst people on the autistic spectrum: A personal exploration of a new social movement based on neurological diversity* [Honours dissertation, University of Technology, Sydney].

2

Neurodiversity—The Words We Use

Caleb—A Missed Opportunity

Ms Bradbury's grade 8 science class was just settling in and getting ready for a lesson on heat and expansion. Ms Bradbury was at her desk taking attendance and completing a few other logistical needs while I stood in front of the class, holding up a copy of their science textbook that happened to have a colorful photograph of the Hot Air Balloon Fiesta in Albuquerque, New Mexico on its front cover. I was married in Albuquerque during the balloon fiesta so I was really excited about making the connection.

I asked the class if anyone knew what made hot air balloons float. I looked around for a response. Most of the students just continued to stare down at their phones, and poke and flirt with one another as junior high schoolers are apt to do. No one took the bait.

One student—I'll call him Caleb—walked continuously around the perimeter of the class, pacing quietly while delicately running his fingers over every surface in his path. He dragged his hand over the window, the bookshelves, and even my face as he walked right in front of me. As Caleb walked close to me, I realized he was mumbling under his breath. He was repeating a rhythmic chant to himself over and over as he circled the room.

*"The fire heats the gas.
The particles spread apart.
The gas expands.
The balloon rises.
The fire heats the gas.
The particles spread apart.
The gas expands.
The balloon rises . . ."*
Which is, by the way, exactly how a hot air balloon works.

Caleb danced his fingers along the wall, singing his little song to himself, while I tried to get his attention. I wanted to coax him into sharing his knowledge with the class. But once I called out his name, Caleb's aide—a paraprofessional who felt it was her job to stop him from disrupting the class—ushered Caleb back to his seat, apologizing to me profusely and hushing him up while she pressed him on his shoulders, forcing him down into the chair.

This confinement, of course, prompted the inevitable meltdown. As Ms Bradbury started the lesson on heat and expansion with the rest of the class, we all practically had to yell to be heard over Caleb's ongoing outburst. The aide finally soothed him as she'd been instructed, with his favorite pen and notebook. He sat for the rest of the class quietly drawing buildings. He was completely disengaged from the conversation on heat and expansion.

Ms Bradbury told me the story of Caleb wasn't unique. She was aware that many of her students knew much more than her lessons could reveal. She knew many of their talents were masked, and even stifled, by the constraints of school. Some of her students couldn't or wouldn't write or speak, so she had no demonstration of their knowledge, at least in the way school outcomes were measured. Some of her students were so exhausted on a test day that they handed in blank sheets of paper void of the brilliance she'd seen the day before. Some were forced to sit in a chair, or take off their headphones, or sit next to their bully—distracting them and sometimes enraging them and possibly burying their talents in STEM problem solving.

Caleb had an Individual Program Plan (IPP)—the Nova Scotian equivalent of what is called an Individual Education Program (IEP) in the US. Caleb had been diagnosed with autism. His IPP provided modified outcome goals that were adapted to his observed capabilities and restrictions. Unfortunately, his IPP did *not* call for rigorous, open-ended scientific explorations where Caleb could have excelled. Since that day, I have asked nearly every educator I've met if they've ever seen the recommendation to engage a neurodivergent learner in a maker space, or in the music room, or the art studio as part of their IEP. To let them learn in an environment where they would excel. Not one teacher I've asked has ever said yes.

My observation of Caleb's experience, and many like him, was the catalyst for an ongoing discussion between Ms Bradbury and me. What if we let kids move around, express themselves as they chose, and encouraged them to become an expert in whatever interested them? What if we let every student's brain work in the way it worked best? What would her classroom look like then?

Children like Caleb have much to offer the world. They aren't broken. Caleb clearly understood the science, perhaps better than many of his peers, but the requirement of sitting in a chair and behaving in an "appropriate classroom manner" excluded him in the lesson. That was a loss, not only for Caleb, but for his classmates who could have learned from him, and for those who will rely upon innovative STEM thinkers, like Caleb, in the future. In other words, for all of us.

Why Words Matter

Nearly everyone I interviewed for this book—neurodivergent learners, their families, and their employers—talked about how much the terminology around neurodiversity matters. The way we talk about students like Caleb matters. How employers and employees talk about their neurodivergent colleagues matters.

I try to weave my way around the delicate balance of language in the book, using different terms in different circumstances.

Most of the people I spoke with who identified as neurodivergent had chosen ways of speaking about themselves. For people who identified as autistic, they most often referred to themselves as an autistic person (in an identity-first fashion), whereas people who identified as having ADHD often used a person-first description, saying they are a person with ADHD. I try my best to preserve the ways people referred to themselves in my writing. I apologize in advance if any of my words make a reader feel discounted or marginalized. That is the opposite of my intent. To discuss the research and practice around neurodiversity, however, it is almost impossible to escape the terminology used in the medical and educational communities.

Categories of Neurodiversity

Much of the current research in psychology, education, and neuroscience uses diagnostic categories to break down neurodiversity. These categories are usually described as a set of deficits, and using these labels has many drawbacks. For example, a diagnosis of *autism* is often prompted by perceived difficulties in areas of social communication (e.g., making eye contact and reading body language) and a tendency towards repetitive and restricted behaviors. Autism is also often associated with high sensitivities or intolerances to noise, touch, and foods. The variations of autism have been commonly seen as a spectrum and referred to as autism spectrum disorder (ASD), ranging from very subtle to very severe.

Attention deficit disorder (with or without the added H for hyperactivity) is typically associated with inattention, distractibility, and low working memory. ADHD adds impulsive and high-risk behavior to the mix. ADD and ADHD are diagnosed through exhibited behaviors that relate to trouble with organization, staying focused, planning, and follow-through. Neuroscience research indicates a relationship between ADHD and low levels of dopamine, the hormone that yields pleasure and reward, which may be the reason for an associated penchant for novelty and high-thrill activity.

Dyslexia may be diagnosed when a learner has trouble learning to read. Dyslexia is caused by a difference in the way the brain translates symbols and sounds into language. *Dyscalculia* has a similar translation issue between symbols and numbers, where numbers, letters, and equations seem disorganized and nonsensical, often described as "swimming on the page." There are several neurological, structural, and functioning differences in the brain that are thought to be related to dyslexia and dyscalculia, none of them related to intelligence or actual literary or mathematical skills.

These types of categorical labels are often used to initiate IEPs for students that include remediation and adaptations to the mainstream curricula. When the student has a specific diagnosis, a plan is made to deal with that set of issues. In current educational systems, these diagnoses and interventions are often necessary for families to access the supports that they feel they need for their neurodivergent children. Many school programs designed to help with communication, reading and writing, and behavioral supports are only available to students with an official diagnosis or assessment.

These diagnostic categories also help researchers to get a handle on the complexity of studying differences in learning and cognition. My team's research proposals are often declined when reviewers want to see a direct connection between an intervention and a particular diagnosis as if learners are siloed into these clear and distinct categories. That's just the way the field has developed. In reality, however, individual diagnoses are often coincident and hard to untangle. Adding even more complexity to the discussion, it is observed that many external physical, emotional, and social stresses can create the same types of changes in the brain that are associated with autism or ADHD. Chronic stress, trauma, or lack of sleep and nutrition can compromise critical brain function, creating neurological differences that affect thinking and learning. And these are all on the rise in today's society. So, the labeling game gets more complex.

Complications With Categorization

Cultural differences may also complicate who is diagnosed with learning differences and how their accommodations are administered. For example, in the US, the prevalence of autism diagnoses is less in communities of color than in Caucasian communities, and children of color are typically diagnosed later than their Caucasian counterparts. Psychology professor Yvette Harris explains that the differences in diagnoses and treatments among children may also be due to cultural differences in the perception of child development. In some cultures, autism is seen as a gift. In other cultures, eye contact is not encouraged between children and adults, so the lack of eye contact that often is an early indicator of autism may go unrecognized or not be seen as a concern. Harris notes that parents' opportunity and financial ability to access diagnostic measures and treatments play a large role in whether their child is diagnosed and treated in a timely fashion.

Many autism advocates say: "If you meet one person with autism, you've met one person with autism," meaning that it is difficult to make generalizations about the huge variations in physical, emotional, and cognitive aspects of each person's autism. The same can be said for other neurological differences, such as ADHD and dyslexia, which are also seen to vary greatly among individuals. This is what makes labeling learners so tricky. Moreover, deficit-based labels often lead to stigmatization, which was a common adversity that plagued many of the neurodiverse people I interviewed for this book. Many people spoke of the feeling that they were "stupid" or "unfit" somehow, and most spent a large part of their lives assuming they couldn't get a "real job." In some cases, this stigma turned into a self-fulfilling prophecy.

Asset-based Perspective of Neurodiversity
When studying learners with characteristics similar to autistic learners, Hans Asperger described a set of behaviors (including pedantic speech content, impairment of two-way interactions, excellent logical abstract thinking, isolated areas of interest, repetitive and stereotyped play, and ignorance of environmental demands), often referred to as Asperger's syndrome, that are thought to more likely to appear in children of high intelligence and superior abilities. Asperger's syndrome, or just Asperger's, is sometimes used to describe people who exhibit some of the aspects of autism while being strong verbal communicators and

often excelling academically in certain areas. Analogous research showed many brain-injured and hyperactive children also exhibited superior intelligence to their peers.

Psychologists Brody and Mills suggested that "descriptions of individuals who are academically talented and individuals who have learning disabilities should be examined and expanded to include students who exhibit the characteristics of both exceptionalities simultaneously in related and unrelated areas." This shift in perspective leads to an examination of the assets as well as the deficits associated with learning differences. An asset-based perspective is more in line with how STEM companies are looking at neurodiversity by seeing people who think differently as a competitive advantage.

The dual nature of the cognitive differences in these learners has led to the term *twice-exceptional* or *2e* used to describe students who have both talents and academic needs outside the norm. Reis and colleagues created the following operational definition:

> Twice-exceptional learners are students who demonstrate the potential for high achievement or creative productivity in one or more domains such as math, science, technology, the social arts, the visual, spatial, or performing arts or other areas of human productivity AND who manifest one or more disabilities as defined by federal or state eligibility criteria. These disabilities include specific learning disabilities; speech and language disorders; emotional/behavioral disorders; physical disabilities; Autism Spectrum Disorders (ASD); or other health impairments, such as Attention Deficit/Hyperactivity Disorder (ADHD). These disabilities and high abilities combine to produce a unique population of students who may fail to demonstrate either high academic performance or specific disabilities. Their gifts may mask their disabilities and their disabilities may mask their gifts.

Dr. Susan Fletcher-Watson, a psychologist and autism researcher, notes that terms like "disorder" are pathologizing, meaning they treat difference as something unhealthy that needs to be fixed. To steer away from a deficit view and stigmatization of learning differences, an Australian sociologist Judy Singer coined the term *neurodiversity* in her graduate work, relating neurological diversity to the diversity movements related to feminism and gay rights. The terms neurodiversity and *neurodivergent* are seen as less judgmental than terms such as *learning disorder* or *learning disabled*.

Neurodiversity advocates see neurological differences as a natural part of human variation, but there is a danger of being overly broad in the definition of neurodiversity. If the term is used to describe everyone, then specific needs and strengths of neurodivergent learners may get lost in the story. Political science professor Dana Lee Baker explains: "expansive definitions of concepts such as neurodiversity run the risk of death by diffusion. No two human brains are identical. To the extent that neurodiversity includes everything, then it might come to mean nothing." Baker notes that finding this balance between inclusion and specificity around neurodiversity presents a challenge for researchers and policymakers, but it is worth tackling.

Terminology that pathologizes differences—words such as disorders, disabilities, or syndromes—places the burden of change on the learner rather than considering how the environment can be changed to help them thrive. Our education systems, by still using the language of disability, may propagate some of the stigmatization that often plagues neurodivergent learners, and may keep them from succeeding in their areas of strength—masking their assets in STEM problem solving, exactly where we need their extraordinary talents. Author Steve Silberman, in his book *Neurotribes*, points out that many of our greatest STEM problem solvers have exhibited evidence of autism, suggesting an overlap between autism and genius. Similar claims have been made about people with ADHD and dyslexia. To leverage this potential overlap, I hope this book can help undo some of the marginalization of neurodivergent thinkers. I hope the unique talents and problem-solving skills associated with neurodiversity will be recognized as potential strengths rather than a disorder that needs repair.

So, what are you thinking?

1. What words are you most familiar with to describe neurodivergent learners? How do those words shape how you think about these learners?
2. What opportunities are presented when language is changed?
3. What might be the risks or limitations of a neurodiversity perspective?

Maria and Paul—Coming to Terms

The hospital waiting room was cheery with primary colors and cartoon characters on the walls. A young mother, Maria, sat nervously by the door while her husband, Paul, tried to fold himself into a chair befitting a five-year-old. They waited silently for the pediatric psychiatrist to return with the results from a recent battery of tests for their son, Felix.

"Your son has autism." The words hit Paul and Maria like a slap in the face. They were stunned. Autism was a word used to describe other children, not their own. Autism meant a child who sits wordlessly in a corner, not interacting with others. That was not their Felix.

Felix talked all the time. In fact, sometimes he just wouldn't shut up. Felix was amusing and bright. He enjoyed playing games with his family. At three-and-a-half-years-old, he could already remember each Candyland card that had already been played. He could rattle them off in sequence and predict which was left yet to come up. He was not yet completely toilet-trained, but he was only three and a half. That's not late, they kept telling themselves, he's still normal. Felix was independent. He just liked things his own way.

Maria had noticed that some of Felix's other behaviors were a bit peculiar compared to other children. He only would eat carrots from the left side of the plate, and only while sitting on a red chair. The other moms at playgroup had a chuckle at his individuality. Sure, he was different, but autism? No way!

But then, Felix still wasn't walking at 21 months. He would sit happily for hours, alone, with a set of colored beads lining them up in sequential order, giving each of them a unique name and voice. He became enraged when Maria tried to scoop the beads up off the floor without respecting their structured order. On an outing to the grocery store, Felix had torn off his clothes in the middle of the produce aisle and refused to put them back on, ending in a naked fit on the floor. As their child further withdrew into his own preoccupations—seemingly more concerned about avoiding rough fabrics, new foods, and lighting choices than receiving affection from his own parents—Maria and Paul had grown increasingly bewildered at how to support the child whom they loved with all their hearts, but with whom they didn't know how to connect.

As the diagnosis of autism sunk in, the doctor visits continued, and clinicians continued to list services available for Felix—mostly consisting of recommended food choices, speech therapy, and clothing textures. Maria and Paul started shifting their expectations. They started seeing themselves as lifelong caregivers, financiers, and advocates for a child who the world saw as broken. They were extremely appreciative of the state-supported services they received and felt fortunate that Paul's income was enough to pay for home support, giving Maria some time to care for their newborn. They don't know what they would have done otherwise. For the next few years, Maria and Paul followed the advice that every medical and education professional gave them. They pursued the therapies, they joined the support groups, and they enrolled Felix in a school that had all the necessary supports for children with autism.

By the time Felix was eight years old, however, Maria and Paul started wondering if they were doing the right thing. While Felix was cooperative and accepting of the many physical and cognitive supports in his school programs, at home Felix had bloomed into an avid reader, he could beat both of his parents at chess, and made his way through the highest levels of a math puzzle game intended for adults. He seemed so different from the other kids his age, so much higher achieving in some areas, yet still needing the structure, sensory adaptations, and social supports his school's autism program provided. When he attended summer camps with "normal" kids, he complained that they all ran too fast, yelled too loudly, and played games that Felix couldn't follow. They called him a baby for still liking to play in the sandbox. Felix retreated into himself when he was around non-autistic kids.

Felix is now transitioning to a large public high school. Maria and Paul are nervous as they watch Felix leave the comforts of his structure and supportive setting, yet they are excited at the opportunities he may have to pursue his interests in rocketry, science fiction, and computer coding. They hope he may have more of a chance to find other kids with his interests and talents because of a larger catchment area of the new school, but they also know it comes with its threat of more bullying and more isolation, as well as the requirement for more independence on his part. They only hope that Felix will be allowed to be Felix, wherever he goes.

When talking with parents about their experiences raising neurodivergent children, I found their stories varied as wildly as their children's. Some suspected a diagnosis before seeking help and some never saw a medical professional at all, receiving supports primarily through school. Some were bewildered by their child's neurodivergence, others had predicted it and prepared, but they all had the same hopes that their child would find their way to being a contributing and accepted member of their community.

What Is Normal?

In a 2015 article for the American Medical Association's Journal of Ethics, Thomas Armstrong wrote:

> In the basement of the Bureau International des Poids et Mesures (BIPM) headquarters in Sevres, France, a suburb of Paris, there lies a piece of metal that has been secured since 1889 in an environmentally controlled chamber under three bell jars. It represents the world standard for

the kilogram, and all other kilo measurements around the world must be compared and calibrated to this one prototype. There is no such standard for the human brain.

Yet, a discussion of neurodiversity begs the question—what is neurotypical? In other words, what is normal?

The Normalization of Education

I spent the first three years of my undergraduate studies at Keene State College in New Hampshire, which was formerly called Keene Normal School. Normal school was the term used to describe a teachers' college, originating from the École Normale founded in 1685 by St. Jean-Baptiste de La Salle of the Catholic Church. The church initiated normal schools with the apparent intent to lift poorer children from the dredges of society by emphasizing conformity and adherence to a common set of societal rules through teacher education and curriculum.

This notion of cultural norms was not limited to the field of education. In the 19th century, French statistician Adolphe Quetelet introduced the idea of averaging human physical features (e.g., height and weight) in an effort to define "the average man." This work is a forbearer to today's measure of body mass index (BMI). The shift towards focusing on a physical ideal placed those furthest from the mean as somehow deviant and needing to be re-centered. The optimization of the human race meant pushing individuals out of the fringes and back towards the normal part of the curve, resulting in a normalization with potentially dangerous consequences. Lennard Davis, a professor of disability studies, points out that this tendency to suppress deviations from the norm provided underpinnings for social and political movements such as eugenics. Davis writes:

> As coded terms to signify skin color—black, African-American, Negro, colored—are largely produced by a society that fails to characterize "white" as a hue rather than an idea, so too the categories "disabled," "handicapped," and "impaired" are products of a society invested in denying the variability of the body.

It was under this social and political landscape of normalization that Austrian-American psychiatrist Leo Kanner conducted research to identify the group of cognitive, social, and emotional conditions now familiarly recognized as the condition of autism. In his book, *Neurotribes*, journalist Steve Silberman gives a fascinating account of how the science of autism and the political views of the last century, including eugenics, have unfolded into our current, still evolving, understanding of neurodiversity. Kanner was operating in a time when children who exhibited neurodivergent behaviors—those considered ill-fitting with the norms—were typically institutionalized and isolated from their families. At nearly the same time, Hans Asperger was studying patients with similar characteristics as people with autism, distinguishing those patients he thought to be "high functioning." These two branches of research have set the stage for the ambiguous and evolving understandings of autism we have today.

What Is Neurotypical?

The term neurodivergent implies that there is such a thing as *neurotypical*. This idea is particularly flawed when one looks to the average to define typical, which is what researchers tend to do. To make this point, in his book *The End of Average*, author Todd Rose summarizes the groundbreaking work of neuroscientist Michael Miller, who imaged the brains of 16 individuals while they were recalling previously heard words to study brain activity during verbal memory tasks. Miller noted that not only were the brain activity images of his research subject all different from one another, they all differed from what was thought of as the "average" brain. Even more outstanding is that these deviations from the so-called average brain were not subtle, they were extensive. People's brains all appeared to work very differently from one another. Since Miller's work, researchers have observed similar differences in brain function in a variety of other types of tasks, including procedural learning and emotion. Rose argues that by aiming for an average brain, we may be missing something because an "average-size-fits-all" model ignores our differences and fails at recognizing talent.

So, this all begs the question: What is neurotypical? To dig deeper into the complexity of this question, let's consider the following three examples.

Example Number One
Images of human brain activity show the most active regions of the brain during language and speech tasks are located in the left hemisphere of the brain. Similarly, regions typically active when interpreting visual information and spatial processing are located in the right hemisphere. This left–right lateralization of the brain has led some to a somewhat oversimplified view of people as being either "left-brained" or "right-brained," but brain lateralization is complicated by left and right-handedness of people. It turns out that for about one third of left-handed people, the active regions during language/speech processing tasks are in the right hemisphere and visual/spatial processing is more active in the left. This finding leads many researchers to eliminate left-handed people from their neuroscience studies so their data wouldn't complicate their results.

So, what are you thinking? Does this mean we should consider only right-handed learners as neurotypical?

Example Number Two
Approximately 20% of students in English-speaking school systems are diagnosed with difficulties in reading. There are many reasons for these difficulties, one being a form of dyslexia where brain has trouble connecting the appearance of letters and sound of syllables with the words that they help form. In English and other alphabetic languages, this connection is a two-step process. Readers need to translate the letters to sounds and combine those sounds into words before their brain then translates the resulting word into its meaning. For many children, that first step of breaking down of the symbolic nature of the alphabetic characters into words is a challenge that impedes their reading. In some languages, such as many Chinese languages, the characters do not represent sounds that are combined to form words, but rather the characters are infused with the meaning (or partial meaning) of the word itself. In studies comparing children learning to read in alphabet-based languages (e.g., English) to

those learning in a non-alphabetic language (e.g., many Chinese languages), this symbolic translation aspect of dyslexia was not observed.

So, what are you thinking? Does this mean that the definition of a neurotypical brain should depend on the language being learned?

Example Number Three

Anita's mother works the late shift on weeknights, so Anita locks the door tight and holds her hands over her little brother's ears so he doesn't wake to the gunshots outside the window throughout the night. Anita sits up with a flashlight, reading the books her mother brings home from the library about faraway lands, outer space, and anywhere that isn't where she is. In the early morning, when her mother finally comes home and crawls into bed with them, Anita can only let herself drift off to sleep for an hour before she must get up for school. Anita's brain is exhausted. It has also become habituated to a state of heightened vigilance. Even as she sits in school where she is told she is safe, she cannot focus on anything but the doorway where she imagines violence might come find her. Barely hearing the teacher's voice as it blends in with all the other normal noises in the room, Anita passes each day in a fight-or-flight-ready mode, never settled enough to focus on anything else. While Anita has an encyclopedic knowledge of the things she reads about in books, her inability to pay attention in the classroom leaves teachers finding her uncommunicative, uninvolved, and showing no evidence of learning.

So, what are you thinking? Does this mean the definition of neurotypical should depend on a learner's sense of safety and comfort?

While it is nearly impossible to define neurotypical, and therefore also impossible to pin down an exact definition of neurodivergence, we still can recognize the value and importance of neurodiversity in STEM problem solving. We can examine the extraordinary talents that many neurodivergent learners exhibit and look ahead to education that embraces and nurtures these differences rather than trying to fix them. I believe we will all be better for it.

Bibliography for Chapter 2

Armstrong, T. (2015). The myth of the normal brain: Embracing neurodiversity. *AMA Journal of Ethics*, *17*(4), 348–352. https://doi.org/10.1001/journalofethics.2015.17.4.msoc1-1504

Asperger, H. (1991). Autistic psychopathy' in childhood. In U. Frith (Ed. and Trans.), *Autism and Asperger syndrome* (pp. 37–92). Cambridge University Press.

Blume, H. (1998). Neurodiversity: On the neurological underpinnings of geekdom. *The Atlantic*. https://www.theatlantic.com/magazine/archive/1998/09/neurodiversity/305909/

Brody, L. E., & Mills, C. J. (1997). Gifted children with learning disabilities: A review of the issues. *Journal of Learning Disabilities*, *30*(3), 282–296.

Cruickshank, W. M., Bentzen, F. A., Ratzeburg, F. H., & Tannhauser, M. T. (1961). *A teaching method for brain-injured and hyperactive children*. Syracuse University Press.

Davis, L. J. (1995). *Enforcing normalcy: Disability, deafness, and the body*. Verso.

Hu, W., Lee, H. L., Zhang, Q., Liu, T., Geng, L. B., Seghier, M. L., Shakeshaft, C., Twomey, T., Green, D. W., Yang, Y. M., & Price, C. J. (2010). Developmental dyslexia in Chinese and English populations: Dissociating the effect of dyslexia from language differences. *Brain*, *133*(6), 1694–1706. https://doi.org/10.1093/brain/awq106. Epub May 20, 2010; PMID: 20488886; PMCID: PMC2877905.

Immordino-Yang, M. H., Darling-Hammond, L., & Krone, C. (2018). *The brain basis for integrated social, emotional, and academic development: How emotions and social relationships drive learning*. Aspen Institute.

Pujol, J., Deus, J., Losilla, J. M., & Capdevila, A. (1999). Cerebral lateralization of language in normal left-handed people studied by functional MRI. *Neurology*, *52*(5), 1038–1038.

Reis, S. M., Baum, S. M., & Burke, E. (2014). An operational definition of twice-exceptional learners: Implications and applications. *Gifted Child Quarterly*, *58*(3), 217–230. https://doi.org/10.1177/0016986214534976

Rose, T. (2016). *The end of average: How to succeed in a world that values sameness*. Penguin Random House UK.

Silberman, S. (2015). *Neurotribes: The legacy of autism and how to think smarter about people who think differently.* Atlantic Books.

Singer, J. (1998). *Odd people in: The birth of community amongst people on the autistic spectrum: A personal exploration of a new social movement based on neurological diversity* [Honours dissertation, University of Technology, Sydney].

3

Neurodiversity—The Competitive Advantage

Dennis—A Journey From Dishwasher to Programmer

Dennis grew up in a sprawling suburb of a large Canadian city and attended public school throughout the 1970s and 80s. While he remembers being pulled out of class for special reading instruction when he was younger, mostly he remembers trying to fly under the radar in school. He was happier not to get any special attention from teachers or peers, or really anyone else at all.

When Dennis was twelve years old, all he wanted for Christmas was an Atari computer. Even then, however, he realized it was an outlandish request for his parents' household budget. He was gifted an old hand-me-down computer from a family friend instead, and he taught himself some basic programming skills. In high school, Dennis had a lot of free time since, as he says, his "social life consisted mostly of dodging abuse from other kids for being a weirdo and mowing the neighbor's lawn." He used the lawn-mowing money to buy more parts, and soon he started to build his own computers.

Dennis couldn't afford the high-end graphics cards required for many of the newer video games, so he learned how to make his own command-line games that he played with others on the Internet using a dial-up modem he salvaged and repaired from the office where his father worked. On the Internet, he was able to play with other people around the world. They didn't know who he was. They didn't act like he was weird. They treated him like he was smart.

Dennis acknowledges that his schooling must have "done him some good" because he is able to read and write "well enough." He can also do pretty complex math calculations in his head without any problem. His recollection of school, however, is a place of shame, pity, and defeat. Something he never wants to experience again. After barely graduating from high school, there was never any conversation of Dennis going to college. It was silently understood that Dennis would continue to live with his parents and get a job washing dishes at a local diner. For over a decade, he says, he was made to feel grateful to be paid minimum wage for cleaning up after people.

Because Dennis continued to live with his parents, however, he was able to save all his wages for a better computer. Even when he had enough money, however, he still chose to build his own because by this time, Dennis told me, he "couldn't buy anything as good as he could make himself." Dennis kept track of every cent he earned and all his expenditures. He gathered and organized all the necessary information to get the best price for all his computer parts. He monitored how long each component survived and tracked the long-term cost benefit of using one product over another. He spent hours poring over the data and macros in his evolving spreadsheets to keep up with cost changes and fluctuations in available parts, but Dennis didn't consider this real work.

In the early days of the World Wide Web, Dennis also learned HTML and scripting languages so that he could build computer games of his own. He shared his games with people he met on the Internet, who loved them and offered to buy them. It still never occurred to Dennis that he could be a professional programmer or game designer. After all, he said, he was just a dishwasher who still lived in his childhood bedroom and who had nearly failed out of high school.

Among many other personal pet projects, Dennis built an extensive website to curate heavy metal hits used in the score for popular movies and video games, with links to the original songs. It started as a basic web page Dennis created out of his own interest after high school, and it grew into an entire fan site for over two decades. As with the meticulous accounting programs Dennis had created for his own computer financing, and the hundreds of computer games he had built for himself, Dennis never publicized his music website. He never shared it with the world because he saw no point.

But the Internet is a funny place. In 2016, Dennis received an email from someone who had stumbled upon his heavy metal web site and tracked him down through some fan forum chats. This emailer was a fan of heavy metal and of ancient video games and was thrilled to have come across all the material on Dennis' site. He was also the CEO at a small tech company in Silicon Valley and the coordinator of a local human resources network seeking innovative talent. He wanted to know about other work Dennis had done.

That interaction eventually landed Dennis at a job fair targeted specifically at neurodivergent thinkers, a term Dennis had never heard before. Dennis was 44 years old. He had never been diagnosed as having autism, ADHD, dyslexia, or any other type of learning difference, but he soon realized that he had many of the indicators. He started with self-diagnosis tools online and then was able to enroll in a research study at a nearby university that came with a free set of diagnostic tests at the local hospital. He also found a group of professional adults with autism who met in a church basement just blocks from his house, something he didn't know existed before.

Soon Dennis was hired by EY, a large consulting firm that goes to great lengths to attract and support neurodiverse talent. He is consulting on a programming team that serves banking clients around the world. Dennis is learning to interact more with clients and colleagues in team meetings, though he still prefers the time when he can work alone on a task until it is perfected. He says he is uncomfortable sharing his work with others until he feels certain all the bugs are gone. He has a rigorous set of tests that must be completed before he feels he can call it complete. Often the debugging programs he writes to make sure his code works perfectly take longer than creating the piece of code itself, but he wants to make sure every condition has been thought of and every possible use case has been tested. Dennis says he sometimes stays up all night to make sure this kind of perfection can happen, even though it makes him very tired in meetings the next day. He says he used to be able to go without sleep for two or three days at a time and still be productive, but that has changed with age.

Dennis wonders if he would have always been this thorough and diligent at work if he hadn't gone through life thinking he was inept. "So maybe it's a good thing in the end," he ponders. He says sometimes he feels he needs to work harder just to earn the trust of his colleagues. He is still afraid he will let them down. But he also says he likes working with a team with others who complement his strengths. It makes him realize he doesn't have to be perfect at everything. While Dennis is not sure he will ever want to be the team member who stands before a client and gives the pitch, or the one who soothes over any personnel issues, he feels confident that he is the guy they can rely upon to get the programming job done right. He will think of everything that could possibly go wrong and make sure none of it happens. He likes his code to be perfect just because that is what gives him personal satisfaction. "It just feels really good when it's right," he says.

As a professional programmer with a professional salary, Dennis has finally found the confidence and resources to move out of his parents' home. He now lives with his partner, whom he loves very much—yet another facet of his life he had previously thought impossible. He had thought he was a loser in life because that is what other people had made him believe.

Neurodiversity in the STEM Workforce

In the early 2000s, STEM companies started taking note of particular strengths among autistic people for certain types of detail-oriented, repetitive tasks. In 2004, a Danish IT specialist named Thorkil Sonne founded a consulting company for software testing and quality control called Specialisterne (Danish for

"The Specialists") where approximately 75% of the employees are autistic. To recruit and hire his staff, Sonne had to create a new way of "interviewing" employees. He wanted to reveal the extraordinary talents of autistic people for software testing tasks while mitigating the social communication barriers that often stand in their way of getting hired.

Specialisterne developed a set of performance tasks to assess potential employees' strengths and weaknesses for software testing. These performance tasks included robotics activities (such as writing a program to move the robot from point A to point B in the least number of moves) and debugging tasks (such as finding an error in a program that somebody else had already written). Specialisterne called this process "trying out" rather than interviewing. They wanted to give potential employees a chance to show what they can do, rather than just talk about it.

Since the start of Specialisterne, numerous global STEM companies including Microsoft, SAP, and EY, have initiated similar neurodiversity hiring programs. In preparation for this book, I started calling around to the directors of these programs asking them what motivated them to increase neurodiversity in their companies. Honestly, I expected to hear something like, "The CEO's nephew has autism," or "A VP is on a foundation board for kids with special needs." I expected there to be some kind of personal connection that motivated a sense of philanthropy. While that may be true in some cases, that is not what I heard from these folks at all. Yes, these companies acknowledge that there are public relations and marketing benefits to being seen as "doing good," and they see improved morale and employees feeling good about working for a company that is "doing the right thing," but again and again I heard the same things from these neurodiversity recruitment specialists—it's all about the talent. Their companies see neurodiversity as a *competitive advantage*. By changing their hiring practices and expectations of what made a "good" employee, many STEM companies realized that not only were they able to broaden their employment pool, but also "some neurodiverse employees turn out to be very, very good at their jobs, better than anyone the company thought they would be able to hire."

In highly complex STEM fields such as cybersecurity, companies seek candidates with highly detailed pattern recognition, a skill that is shown in research to be prevalent with autism. Austin notes the anomalous signals in reams of data that may show a cybersecurity breach can only be located by an exacting human brain. Dr. Andrew Begel, an autism researcher formerly at Microsoft, told me in conversation that good software engineers are often systematic, methodical, and meticulous visual thinkers. They may enjoy dwelling in the minute details of a problem. Begel explained: "That is where many autistic people want to spend their time. They often opt for difficulty on purpose. They may intentionally select hard and complex tasks to keep themselves challenged at that detailed level."

As companies increased the neurodiversity of their workforce, they noted clear advantages and thus started to broaden their programs from a focus on autism to a larger pool of neurodivergent problem solvers, including people with ADHD, dyslexia, and the multitude of other differences that often fall under this heading. STEM companies are not only interested in detail-oriented workers for cybersecurity, programming, and quality assurance, they are also seeking creative people with a fascination with new technology and an eagerness to introduce new and disruptive ideas. In STEM companies' quest for constant innovation, neurodiversity of their workforce enables them to include people and ideas from "the edges." Divergent thinking can help companies think "outside the box," which can lead to places they wouldn't have gotten otherwise.

In my conversation with Tammy Morris, the Neurodiversity Center of Excellence Network Leader for EY Canada, she explained: "EY is a company of high performers—any type of obstacle or barrier results in a conduit of innovation. We want people who are creative and ingenious in overcoming barriers, and that leads us to neurodiversity." Companies like EY are tapping into what they see as an underutilized talent source by reducing barriers of entry so that they can get the most qualified team for a job. Harnessing the creativity, systems thinking, and persistence of neurodiversity has proven so successful for STEM companies that they are changing the landscape of the workforce. One of the ironic challenges they are now facing is finding enough neurodivergent applicants. Because a professional career has not seemed possible to many learners who have struggled in school, or because the system that educated them didn't provide them with the confidence and agency to find a career, many potential employee candidates may consider themselves unemployable so they are not looking to be hired.

To get around this, some companies work with colleges and high schools to set up nontraditional work experience programs for neurodiverse populations, particularly in the areas of video gaming, robotic programming, and other activities. EY and other companies hold targeted job fairs for neurodiversity, often coordinating with college and university offices of student services. Nearly every employer I spoke with said their neurodivergent employees were often their most productive employees. They only wish they had more like them.

So, what are you thinking?

1. What talents of neurodivergent learners have you observed that would be beneficial in a STEM problem-solving environment?
2. What types of STEM problem-solving activities do you see that might benefit from neurodiversity?

Sara—A Journey to Genius

My own graduate studies were in astrophysics, and still to this date I teach a university course on life in the universe. So when I was given a chance to interview Dr. Sara Seager for this book, I was a bit starstruck. She is a rock star in my field—a professor at MIT, a MacArthur "Genius" Fellow, and the leader of several important NASA missions for the search for extraterrestrial life. She even has an equation named after her—the Seager equation—which is used to estimate the number of habitable planets in the galaxy. She is also neurodivergent.

She was an adult before she realized she is autistic, though she always felt she acted differently somehow than people expected of her. When Sara was a child, she found it hard to make friends in part because she was always intensely focused on one thing, to the exclusion of others. She found this made it hard to find things to talk about with other people. Small talk is something she had to train herself to do, for other people's sake. She says she'd just as well do without it, but she doesn't want to be rude. Though, she adds, she loves talking with dogs.

Sara's celebrated success doesn't seem to be of much importance to her. She told me in our conversation that she likes the awards because they help her do more work and work with other outstanding scientists. The recognition doesn't motivate her as much as the science itself. She is a persistent, curious, systematic, creative problem solver.

Sara senses that her brain runs faster than most people around her, and while that means she can get a tremendous amount of work done in a short amount of time, she also has to remind herself to put the brakes on occasionally to give others time to catch up. She sees her brain's capacity for idea generation as a positive aspect of her neurodivergence. When she is working on a problem—especially when she can avoid the minutia of daily life—Sara says she has the space to make connections across all the information she knows and new information she's learned. She can put ideas together to innovate new ideas. She explains:

> Creativity in science is not just about choosing a great idea, but also about figuring out how to solve the problem. To make advances in STEM, you have to have that light bulb go off in your head. That comes from intuition built over time. And my brain is good at putting all that information together to solve new problems.

Having attended Montessori school near Toronto for the first few years of her education, Sara loved the open-ended nature of learning where she could dwell as long as she liked in her favorite activity and wasn't forced to follow someone else's plan. Therefore, she was taken aback when she was moved to a public school in grade 3, where she found the class highly structured and focused on teacher-led lessons. As an inquisitive third grader, Sara was annoyed to be told to do what to do rather than to be allowed to follow her own curiosity. After attending Jarvis Collegiate Institute, a 200-year-old public high school known for its outstanding science education, Sara entered the University of Toronto as she said in an interview with Smithsonian Magazine, "with the idealistic view that anything and everything could be described by a physics equation." Since then, pursuing huge questions about the nature of the cosmos and how to discover lifeforms on other worlds provides the wide-open cognitive spaces Sara craves for her generative mind.

While her father had encouraged her to become a physician, Sara followed her own line of inquiry in graduate school and beyond. She finally was given the liberty to take one subject, astrophysics, and explore it to its fullest detail. "That's what I had been wanting to do all along," she says. "I had a good education, and I suppose all those other classes were useful to help make me a well-rounded thinker and all, but I only really cared about the courses that helped me answer the questions I was thinking about at the time."

When I asked Sara if she would change anything about how her brain works, she immediately replied with a firm "no." She attributes her neurodiversity as the reason she can get so much done. Sara says the way her brain can remain hyper-focused on a problem allows her to remain in a state of flow, intently focused on a task for hours at a time, and this comes easily to her. It is her place of comfort. "I get so much done in a small amount of time. I wouldn't want it any other way," she says.

Sara says that to collaborate well with colleagues and to be an effective teacher, she has had to develop explicit strategies for communication. She says she has to remember to give others time to form their own thoughts, so she doesn't come across as rude or insensitive. Sara was about 11 when she started to realize that body language was key to understanding other people's perspectives. Not being able to figure it out on her own, Sara went to the library and read every book she could on it. As a scientist, she took the scholarly route to learning social behaviors.

Sara says she's had to manage her own expectations and learn how to make small talk since that is what helps others feel comfortable. She and a neurodivergent colleague created a guidebook for themselves to help them work with other scientists and students. The book included cues to remember in different social situations such as a reminder to count to ten to give others a chance to speak. Their guidebook also includes a list of topics most people are interested in—such as dogs, kids, and the weather—though the small talk is of little interest to her. She and her colleague used this guide to help learn how to interact the way others expected her to, how to act in a way that others thought was "normal."

Sara sometimes wonders why it was the neurodivergent people who had to create the guidelines book—why there isn't a guidebook for others to learn how to work with people like her. Pragmatically, she suggests it's likely because majority rules. "That's the way the world works," she says flatly. "I just wish I got the guidebook at a younger age. I would have liked to know earlier how I was expected to act."

The biggest thing Sara says she would change about neurodiversity is not the differences themselves, but rather other people's perspectives about neurodiversity. She'd like to see neurodivergent children treated differently in school to remove the stigmatization. She credits her father for her experience of growing up relatively unscathed by this stigmatization because he made her feel loved despite her quirky intelligence. Sara fears many kids aren't as fortunate and get left behind, noting that "Students can't reach their potential when shame is attached."

Bibliography for Chapter 3

Austin, R. D., & Pisano, G. P. (2017). Neurodiversity as a competitive advantage. *Harvard Business Review, 95*(3), 96–103.

Bernick, M., & Holden, R. (2018). *The autism job club: The neurodiverse workforce in the new normal of employment.* Simon and Schuster.

Seager, S. (2020). *The smallest lights in the universe: A memoir.* Penguin Random House Canada.

4

Learning in the Brain

Chase—A Sentry at the Door

I met Chase in 2016 when I was working at Hillside Junior High. We were starting project-based learning with the students, and a few of the grade 9 students wanted to make campaign videos like the ones they'd been seeing on social media. Trying to keep some bounds on what was appropriate for a school project, we steered the pair towards the video-editing software in the computer lab and kept a close watch.

I hadn't gotten to know Chase very well because he was often truant and seemingly disengaged when he was in class. When he was present, he usually sat in the back with his hoodie up over his head and slumped back in his chair, as if he was trying to meld into the plastic itself. I assumed he was often asleep. He looked frail and unwell, as if from an overall lack of nutrition, and he appeared unreachable. Completely detached.

One day, when the class was in the computer lab, Chase was working on his own because his partner was absent. I took the opportunity to sit next to him and try to start a conversation. Chase offered monosyllabic answers to my questions, at best, and his eyes kept darting around at anywhere but at me. I had no way to connect with him, so I just sat silently and watched what he did. His technical skills were solid. He fluently navigated the different software packages to do exactly what he wanted to do, which was to make Bernie Sanders fly through the air, interact with several rap artists and cartoon characters along the way, and finally explode into a visual and auditory crescendo at the end.

Sitting that close to Chase, and observing him more carefully, I saw something different in his behaviors than I had noticed in class. The best word I can use to describe him is skittish. While trying to remain tough on the outside, his body language revealed a scared little boy inside. Any unexpected motion or noise sent his limbs into a visceral retreat. His sharp reflexes were jerky and protective. His shoulders hunched and spine curved inward as if shielding his innards from vulnerability. I could watch a shell curl up over him each time another person walked anywhere near him.

Over the next few weeks back in the classroom, I kept a closer eye on Chase. He still slumped into his chair in the back row, hoodie up over his head, but now I saw that he was clearly not sleeping. Far from it. Chase spent every minute of class with his crystal blue eyes darting out from the small opening of his hood, bouncing his gaze from the windows to the door and back to the window, only interrupted with an instant response to any sharp noise around him. Once I noticed his behavior, it was so distinct I did't know how I hadn't seen it before.

When Ms Bradbury tried to corral Chase's attention, asking him a question about the class discussion, he bristled and shut right down, refusing to participate. Refusing to be distracted from his vigilant watch. The computer screen seemed to be the only place that Chase let down his guard. For some reason, he could get lost when he was working there. The video clips he curated were his reality. He was absorbed in stitching them together, getting the characters' motions exactly timed with the background beats he'd selected.

It was clear Chase had the capability to do quality work, yet when it came time for the partners to present the project to the class, Chase was silent. The assessment worksheet he was supposed to complete about the project was left blank on the floor with a dirty boot print stamped on it. Chase could do the project, but he was not willing or able to play the game of worksheets and presentations—school's rigid expectations—that went along with it. He failed the class.

Chase's other teachers had similar reports. He was seen reading novels in the hallways, but never participated in English class. In science class, he had helped design a water filtration system, using YouTube instruction videos as a guide, but then he never turned in the lab report or participated in class discussion. The only place where his grades were okay was in art class, where he was not required to follow instructions. He produced outstanding expressions of his world through drawings, cartoons, and graphic novels.

Before that school year was out, Chase was removed from his home and placed into foster care in another part of the city. I never saw him again. But I did learn a lot more about his childhood after he left. Chase had been beaten frequently as a child, often coming to elementary school with bruises and soiled pants. He lived in a car with his father for a while and had since lost any familial support as his mother and aunt succumbed to addiction. He ended up staying with his uncle, as he had for a month or two at a time throughout his life, who apparently had beaten him regularly. Chase ended up lost in the system. I will never forget his darting eyes.

The year after I worked with Chase, I learned the term hypervigilance. It refers to an elevated state of responsiveness in the presence of a perceived threat. In Chase's brain, after all the trauma he'd been exposed to, he perceived a threat was more likely to come from the outside than from his teacher talking in the front of the class. So that's where he paid all his attention—a sentry of the windows and the doors.

Brain Structure and Function

If I had a chance to do it all over again, I think I'd become a neuroscientist. The brain fascinates me. It feels like the newest frontier for investigation, and we now have tools to learn so much more than we ever could before. Many of the most exciting questions about neurodiversity seem on the verge of being answered. But alas, neuroscience is not my area of expertise. I have had to basically take a self-directed crash course in neurology to write this section of the book. I still can't remember all the different names of all the parts of the brain, but I do think I have one very important takeaway from my recent research: The brain is a complex system with many different regions and different networks passing information within and across these regions. While most humans have similar overall brain structure, meaning the regions are roughly the same size, shape, and location as other humans, the differences in the brain's functioning—the way information is transmitted through networks within and across these regions—may be at the heart of what we need to understand about learning.

Neuroimaging
Most of what was known about neurology before the last few decades came from autopsies and early forms of X-ray imaging. Not only did that greatly limit the sample populations (primarily to dead people) or put patients in danger, it also meant that most of what could be studied was limited to the structure of the brain, not its function. Functional Magnetic Resonance Imaging (fMRI) changed all that in the late 20th century.

Educational neuroscientists use fMRI and other imaging methods visualize the electrical and chemical signals that travel between different regions of the brain, and/or the blood flow associated with those signals. This allows visualization not only the structure of the brain but also the transmission of information (via electrical signals) within and across the various regions of the brain. In an fMRI, the person is lying within a machine when mapping is being done, so activities are limited to largely sedentary tasks such as passively reading a passage of text. Headsets such as electroencephalograms (EEGs) and functional near-infrared spectroscopy (fNIRS) allows neuroimaging that, while less accurate, can be conducted in more natural settings.

Structure of the Brain
Within the complex machine of the human brain, the primary component involved with human learning and cognition is the *cerebrum*. The cerebrum has two hemispheres (left and right), which each contain separate lobes responsible for different parts

of brain activity associated with learning. The hemispheres of the brain are made up of lobes that are responsible for different types of information processing. The different lobes across the two hemispheres of the cerebrum must all work together to make sense of incoming information.

The *cortex* of the brain is the medium through which that information is transmitted from one part of the brain to another. The tissue that surrounds the cerebral lobes is called the *outer cortex* and is made up of grey cells called *neurons*. This tissue is sometimes referred to as the grey matter of the brain. Neurons in the grey matter exchange sensory, cognitive, and emotional information with other neurons via electrical and chemical signals. Similar tissue (called white matter) lies underneath the grey matter in the *inner cortex*, which contains the *neural pathways* through which the electrical and chemical signals are sent. These neural pathways within the white matter of the brain serve as tunnels or conduits where information can flow.

Networks in the Brain

Neuroscientists have identified three primary networks that help coordinate activity and information exchange across the different regions of the brain. The *executive control network* coordinates the functions for goal-oriented tasks, such as self-regulation, and attention to keep on task and avoid distractions. The *default mode network* deals with information in the abstract. It is responsible for complex conceptual understanding, creativity, inspiration, and identity development. The *salience network* helps coordinate between the executive control and default networks, using external signals such as hunger, pain, or anxiety to influence what tasks have crucial importance and what doesn't need immediate attention.

The activity of the executive control network is often referred to as *executive function*. Executive function is fundamental to learning as it is important for controlling impulses, sorting out salient information from irrelevant distractions, storing and retrieving critical information, shifting strategies as appropriate to accomplish the task, and maintaining and monitoring progress towards goals. Executive function will be explored in greater detail in the next section.

The default network of the brain helps connect new information with other experiences and ideas you many have so that you can build a deeper conceptual understanding of how your brain works. The activity of the default network is also helping you reflect on how this new information may help interpret an event that happened with your students or a group of friends. The default network also uses this information to help you make predictions about how people may behave in the future. Neuroimaging shows that activity of the default network is greater during higher-order cognitive tasks, such as building conceptual knowledge and divergent thinking.

There is also evidence that activity in the default network decreases during tasks when high executive function is needed. In other words, the activity in one network is reduced to make way for the other. For optimal learning, these networks must operate smoothly together, fluidly exchanging resources among networks as needed. The salience network serves as a gatekeeper to allow the executive and default network to share these neural resources.

These networks are composed of neurons and fibers in the brain's white matter, and are organized in columns and mini-columns, where a mini-column is the smallest component capable of information processing. Variation in the width and length of these mini-columns and how the fine-grained structure of these networks impacts learning is an active area of research. There is some evidence that mini-columns may be spaced differently in autistic and dyslexic brains, and also that these differences may vary with age, with some evidence that autistic brains may have poorer connectivity within the brain, and other research showing higher connectivity in local circuits associated with autism. These neurological differences may explain some of the extraordinary focus and abstraction talents of these neurodivergent thinking, but this research is still young and needs further study.

The Limbic System

Deep underneath the inner cortex lie several glands that make up the limbic system, which are responsible for basic emotions and needs. The limbic system includes the hippocampus and amygdala. The *hippocampus* is responsible for making connections between senses and memory, for example, associating the smell of certain foods with childhood memories. The hippocampus is also important for spatial orientation and our ability to navigate the world. The *amygdala* plays a central role in our emotional responses, including feelings like pleasure, fear, anxiety, and anger, and connecting those emotions to new information. Importantly, the amygdala also regulates what is called the "fight-or-flight" response. When a threat is detected, the amygdala initiates a set of involuntary responses to prepare the body for a fight-or-flight response to the danger.

The fight-or-flight behaviors developed in early human brains in response to danger. If the brain has to prepare to fight a prey or flee from a predator, it sends surges of chemicals, such as adrenaline, throughout the body when a threat is perceived. This chemical response causes increased heart rate and the dilation of arteries and veins to pass blood more quickly, which is appropriate and useful when a person is truly in danger. In this fight-or-flight state, all energy is dedicated to the fixation on a threat. In today's world, that little jolt of energy can be useful in providing the excitement and nerves that often aid our performance. The rush we get before a public speaking or athletic event can help us rise to our potential. We may even voluntarily subject ourselves to this condition for fun if we go watch a horror movie or read a thriller. But to stay healthy, the body needs to come down from the fight-or-flight state and be in a restorative state to heal. Getting stuck in a fight-or-flight state draws resources from higher-order cognitive processes, limiting the opportunities for learning. This can be the result of PTSD and other trauma-related conditions. Sometimes the system doesn't know how to let go. This is when hypervigilance, like what Chase was experiencing, and other overcompensations can occur.

Neuroplasticity

There is growing evidence that the structures and pathways in the brain are not permanent or "hard-wired," and that, in most cases, the brain is somewhat malleable. *Neuroplasticity* refers to the brain's ability to change, reorganize, and grow new neural networks. As we develop, our brain builds and prunes these networks depending on how they are used to optimize the brain for what it perceives will be future usage. Brains can be "exercised" by learning a language, learning to play a musical instrument, and playing puzzles and games, all of which are thought to promote neuroplasticity. This is somewhat analogous to how we build muscles through repetitive exercise that hones the body for similar activities, while others may atrophy if unused. Therefore, just like physical exercise, it is important to keep up with the types of activities you want your brain to be able to continue doing.

Executive Function

The activity of the executive control network is commonly referred to as *executive function* (or *executive functions* or *executive functioning*). In daily life, nearly everything we do requires executive function, from doing laundry to making a trip through the grocery store. STEM problem-solving behaviors associated with executive function include planning, implementation, monitoring, and evaluation of a task—processes required, for example, to develop a new software program, find an error in a spreadsheet calculation, or analyze data from an experiment. So, it should come to no surprise that executive function is at the heart of learning and cognition. In fact, I had an epiphany about my 20+-year career in education after only a few visits to Hillside Junior High. If I didn't do anything to address the executive function needs of these kids, I wasn't going to be able to teach them any STEM. That changed everything.

Executive Function Inside the Brain

From a neuroscience perspective, executive function is made up of three primary components: working memory, cognitive flexibility, and inhibitory control. *Working memory* is the storage of information for the brief time required to process and manipulate it. This is somewhat analogous to the RAM in a computer. It is different from long-term storage, where memories are made and later retrieved. *Cognitive flexibility* is the reshaping of ideas and practices based on new information. This can manifest itself in rule-shifting (the ability to respond to a change in rules) or changing ideas in the face of new information. *Inhibitory control* is the regulation of impulses, emotions, and attentions. This is what provides the ability to think before acting. Inhibitory control is a key component of self-regulation.

To illustrate each of these components of executive function, consider what your brain is doing at this moment just to read this sentence. You are using your working memory to keep the first half of the sentence in mind while you are still parsing out the words in the second half of the sentence. You are using cognitive flexibility to change your existing ideas in light of new information on the page. Finally, if you get bored, you may exercise inhibitory control to keep yourself from skipping ahead or leaving the book altogether if you are distracted by the smell of someone making something yummy in the kitchen.

Executive Function in the Classroom

Compromised executive function, be it from chronic stress, trauma, or structural differences in the brain, is responsible for a large component of neurodiversity in school. This involves students who, in a hypervigilant state like Chase in the earlier story, focus all their neural resources on protection, leaving little left for inquiry, imagination, analysis, and retention. His brain, when in that state, was no longer taking in anything I said. Learners in this state may appear disengaged or scattered because they have trouble focusing their attention, and they may exhibit low working executive function because they are so focused on their threat they do not effectively process and retain new information. To re-engage these learners, more productive neural responses need to be re-employed and re-practiced.

Differences in executive function can manifest themselves, and thus be supported in a wide variety of ways in the classroom. Psychologist Clancy Blair notes that these variations suggest that education should focus on supporting executive function to benefit *all* learners: "If [executive function] can be compromised under conditions of disadvantage, both developmentally and at a given point in time, then [executive function] should equally well be supported under favorable conditions." Blair suggests that identifying the reasons for variation in executive function, as well as the times in a child's development when there may be peak variation among children, will point to the greatest opportunities for making a difference.

The National Center for Learning Disabilities (NCLD) suggests strategies to support executive function including problem-solving approaches such as taking it step by step, relying on visual aids and organizing tools, multiple representations, and planning for transitions. Many of its recommendations are in the area of time management, including use of alarms, scheduling tools, checklists with time estimates, scheduled breaks, and calendars. For general organization skills, NCDL suggests cutting clutter, having separate work areas for different activities, and having scheduled cleaning and tidying times. Finally, the NCLD recommends seeking help from teachers and coaches to support executive function within the learning process.

Executive function differences were rampant at Hillside Junior High. This was epitomized by the frequent task of placing a new worksheet in their class binder, a tool required in Ms Bradbury's classes. For her grade 7 students, organizing a binder was a monumental feat, especially at the beginning of the school year, and throughout the year, each new worksheet was a challenge. While some students could easily open the binder to a new page, wait patiently for their turn to use the three-hole punch tool, and place the correct (and undamaged) page in the appropriate spot in the binder, for others this one activity could take most of a class period or never happen at all. While seemingly simple, this presented obstacles to progress for so many, and pointed right to their difficulties in executing an instruction.

To help guide her students, Ms Bradbury spent the first few months of grade 7 modeling the types of planning and organization behaviors that might help her students. She showed students how she had organized her own binders and how she used them to retrieve information, such as the seating chart and grade book for each of her classes. She repeated this modelling daily, and she also showed them pictures of what an organized student desk might look like with everything in a retrievable place. She let students compare the photos to their own desk and modify their setup. This intentional executive function activity took time away from science lessons, but provided a foundation that fostered more effective and autonomous learning practices later in the year (and beyond).

Providing supports for executive function during problem solving often is proven useful for all students, and they should be no more stigmatizing than providing eyeglasses for reading. Just like I need eyeglasses to be able to read a book, a student may need a highlighting tool to help distinguish salient words from distracters, or another student might need a graphical organizer to keep their ideas easily retrievable. By supporting executive function alongside STEM teaching and learning activities, we don't just provide more inclusive and equitable access to STEM learning; we help unleash the other STEM problem-solving talents that could be hidden under unrelated executive function challenges.

So, what are you thinking?

1. Under what conditions do you feel your own executive function is compromised?
2. What types of supports are useful to help you or your students or employees with executive function?

Kiera and Kelly—Two Different Problem Solvers

Kiera and Kelly were close friends in Ms Bradbury's grade 8 science class at Hillside. They had grown up together in the same housing community and they worked together in class whenever they were given a chance. They shared many common stories, laughed instantly at each other's jokes, and completed each other's sentences at times. They were a team.

Sometimes in class, Kiera had difficulty paying attention to a task at hand and regulating her impulses. She easily got distracted from the classwork. Kiera also had difficulty remembering the explicit instructions of how to complete a task while at the same time attempting the task itself. For example, when working on a math problem, she struggled to focus on the logistics of breaking down the components of the word problem and lining up the equations on the paper like the teacher had showed her, and then she had nothing left in the tank when it came time to reflect about how to use an appropriate mathematical strategy. Juggling her executive function needs at the same time as learning new content was a constant working memory overload for Kiera.

Kelly, on the other hand, had ample working memory to keep a larger goal in mind while sequencing and conducting the individual sub-tasks towards that goal. Kelly didn't have to reread the instructions numerous times. She was easily able to craft and follow a plan. She aced all her tests. Kelly appeared to be the far superior student in class.

I observed Kelly and Kiera when they were working together to create a water filtration system in their science class. Kelly had found a set of instructions online and was carefully listing them one by one on a worksheet Ms Bradbury had provided. Keira was fiddling with some pipe cleaners and plastic wrap and had made a goofy cartoon character while Kelly was getting organized. At one point, however, Kelly's face fell when she realized they didn't have the bolts required for the design she had been so diligently following. Kiera asked a few questions. What were they for? Where did they go? Kiera became interested in the problem while Kelly sunk into further disappointment.

As Kelly struggled to follow the exact instructions provided step by step with the materials they had, Kiera scouted through the equipment bins and devised an alternate method of securing the two filter components together that, in the long run, made the entire task more efficient. Kiera saw the problem differently and didn't get stuck in a way of thinking. She saw the big picture level and generalized the problem into an algorithm, making connections to other types of problems. While Kelly was great at getting organized and following instructions to complete an individual task, Kiera abstracted the problems and devised a strategy to complete the whole category of similar tasks.

Both Kelly and Kiera bring strengths to the problem-solving endeavor, but they are different strengths. It is important to look out for what each learner can bring to the situation and provide opportunities to let those talents shine.

Bibliography for Chapter 4

Blair, C. (2016). Developmental science and executive function. *Current Directions in Psychological Science, 25*(1), 3–7.

Blair, C., & Razza, R. P. (2007). Relating effortful control, executive function, and false belief understanding to emerging math and literacy ability in kindergarten. *Child Development, 78*(2), 647–663.

Buckner, R. L. (2013). The brain's default network: Origins and implications for the study of psychosis. *Dialogues in Clinical Neuroscience, 15*(3), 351.

Bull, R., & Scerif, G. (2001). Executive functioning as a predictor of children's mathematics ability: Inhibition, switching, and working memory. *Developmental Neuropsychology, 19*(3), 273–293.

Costandi, M. (2016). *Neuroplasticity*. MIT Press.

Diamond, A., & Ling, D. S. (2020). Review of the evidence on, and fundamental questions about, efforts to improve executive functions, including working memory. In J. M. Novick, M. F. Bunting, M. R. Dougherty, & R. W. Engle (Eds.), *Cognitive and working memory training: Perspectives from psychology, neuroscience, and human development* (pp. 143–431). Oxford University Press. https://doi.org/10.1093/oso/9780199974467.003.0008

Fung, L. K. (Ed.). (2021). *Neurodiversity: From phenomenology to neurobiology and enhancing technologies*. American Psychiatric Publishing.

Meltzer, L. (2010). *Promoting executive function in the classroom*. Guilford Press.

Meltzer, L. (Ed.). (2018). *Executive function in education: From theory to practice*. Guilford Press.

Mountcastle, V. B. (1957). Modality and topographic properties of single neurons of cat's somatic sensory cortex. *Journal of Neurophysiology, 20*(4), 408–434.

Pereira, A. M., Campos, B. M., Coan, A. C., Pegoraro, L. F., De Rezende, T. J., Obeso, I., Dalgalarrondo, P., Da Costa, J. C., Dreher, J. C., & Cendes, F. (2018). Differences in cortical structure and functional MRI connectivity in high functioning autism. *Frontiers in Neurology, 9*, 539.

Thomas, M. S., Ansari, D., & Knowland, V. C. (2019). Annual research review: Educational neuroscience: Progress and prospects. *Journal of Child Psychology and Psychiatry, 60*(4), 477–492.

5

Learning in the Classroom

Uni High School—Learning About Learning

In the 1980s, I taught in a laboratory high school associated with the University of Illinois. I was fresh out of graduate school for astrophysics without a single education course under my belt, so I had no preconceived notions about how I was supposed to teach or how students were supposed to learn. I taught them the only way that made sense to me. I reflected on my own learning experiences, and I figured I learned much more in my labs, when I got to play around with stuff, than I did sitting at a desk watching someone talk at the chalkboard. Even in math, I had to work out the problems for myself to learn the material—watching someone else do it only got me so far. I also hated being in a stuffy building when it was beautiful outside.

So, during the sunny mid-September week when my first classes started a unit on kinematics (the relationship among position, speed, and acceleration), I brought them all outside and had them run races on the sidewalk. With stopwatches and tape to mark off intervals, the students worked together to devise their own way to measure who was going the fastest at different points in the race, and how to represent that in their reporting. They all understood the basic concept of running a race and taking split times, but in that lesson they were forced to devise with the concepts of instantaneous and average speed, acceleration, and rate of change, even though they used different words. I never had to "teach" a thing, I just had to translate.

I did a similar thing when I wanted the class to understand the inverse square law. This is the law that explains how a force such as gravity is distributed spherically from a point source. The strength of the force decreases as the inverse square of the distance from the center of the sphere (denoted as $1/R^2$). This is also how light is distributed from a light source and basically anything else that spreads out like a sphere from a point. I tried a multitude of drawings and stories, even trying to conjure up a thought experiment of a perforated ball spewing pudding all over the room spherically to give students a picture of it, but I could tell it still wasn't working for them. So, finally I set up a bare 100 W light bulb in the front of a darkened classroom, so the bulb (roughly a point) was the only source of light in the room. I gave each student a 2cm x 2cm piece of black paper. I gave them access to a bunch of measurement devices (light meters and measuring sticks) and told them they could work together but by the end of the class period I wanted them to tell me how much energy was absorbed by their piece of paper each second. And they did it.

I was very, very lucky in that first classroom experience. I see that now. By miraculously good fortune, I was dropped into a community of amazingly bright students in a school supported by passionate and creative educators, administrators, and parents. They let me learn how to teach in the same way I learned STEM. They let me bring my own experiences and ways of thinking to the problem, and I figure it out along the way. Once I got to TERC, I realized that there was an entire field of researchers trying to tap into this pursuit of authentic teaching and learning. There were psychologists and learning scientists who made a career of studying these methods. When I started teaching, I didn't have the words for the type of student-centered learning that came naturally to me. Now I do.

Constructing Knowledge

I now know that the type of learning that came naturally to me as a teacher was grounded in the theory of *constructivism*—the idea that learners do not come to a learning experience as a blank slate but rather construct knowledge by integrating new information into their previously built understandings. Constructivism has deep roots in educational philosophy and cognitive psychology.

Roots of Constructivism

John Dewey, a prominent philosopher and educational reformer of the early 20th century, emphasized that learners are the creators of their own knowledge about the world around them, and that this knowledge is constructed through learners' interactions with their physical and social environments. Dewey argued that each new experience contributes to the development of a learner's mental model or schema of a concept, which deepens as the learner incorporates the new information into the evolving model.

Jean Piaget took a developmental approach to constructivism, believing that children proceed chronologically through a series of discovery phases. Piaget argued that all children went through each phase at the same stage of development, in order and at approximately the same age. Jerome Bruner, however, argued that any child can construct conceptual knowledge at any age if provided with adequate supports. Decades later, Sir Ken Robinson also discarded this rigid developmental perspective of knowledge construction, arguing against how schools sorted children by their manufacturing date, meaning their birthdate. It would be much better, he suggested, to group learners by common interests and complementary skill sets, which often happens organically in real-life situations.

Social Constructivism

The social constructivist, Lev Vygotsky, added the importance of teachers, peers, learning activities, and the learning environment to the knowledge construction process. Learning does not take place in a vacuum but is rather facilitated and mediated by the circumstances of the learning experience itself. This social model of constructivism shifts the responsibility of knowledge building from the individual learner to a larger distributed network—a network that includes the learner, teacher, peers, and the instructional designers of the learning experiences. Teachers, peers, and rich educational materials are all heavy influencers in the learning, making each learning experience unique for each child.

Vygotsky also introduced the very important concept of the Zone of Proximal Development (ZPD), which I will discuss several times throughout this book. The ZPD describes the region of development that lies between what is already within the capabilities of the learner (what they can do currently without help) and what they could do further if provided with appropriate scaffolding and support. The ZPD is critical for most social learning models. It is the idea of "pushing" or "challenging" a learner to take the next step, but also being there with whatever they need to bridge the gap successfully.

Social constructivists sometimes describe learning as being situated within, and thus inseparable from, the learning experience itself. As a naval researcher and an avid yacht racer, the cognitive anthropologist Edwin Hutchins was fascinated by the implicit distributed learning process of a naval crew as they navigate a navy vessel into harbor. He noted that this task would not be feasible by any one individual, as information is needed from all sides of the ship at once, and thus the task required a distributed learning system. Hutchins also distinguished between the technological components of the system, such as navigational tools, and the human cognition of the team of sailors, noting how they were both critical parts of the network. He described these socio-technical learning networks as having cognitive properties of their own, creating a whole that is bigger than the sum of its parts.

Another social constructivist learning model that has had a strong impact on my work over the years is the *cognitive apprenticeship model* described by Allan Collins and his colleagues. This model uses the analogy of a craftsperson passing down a practice to an apprentice, where knowledge building is the craft. In a cognitive apprenticeship model, the teacher is an expert craftsperson of cognition and the student is their apprentice in the learning process. The word teacher is used loosely here because it may be a classroom teacher, or it could be a parent, peer, or anyone else who can play that master role.

The cognitive apprenticeship model outlines six basic steps of teaching and learning. In the first step, the teacher demonstrates the task so that the learner can begin to form a conceptual model for how it can be done. For example, the teacher might solve a typical problem as an example, explicitly explaining the steps as they work through it. Next, the teacher provides coaching by observing while the learner attempts the task, during which the teacher provides real-time feedback that supports the learners' development. The teacher might also structure the task to keep it within the learner's ZPD. For example, the teacher might choose a problem that is close to one they've done previously but also has enough difference to present a further challenge for the student. The third step of the cognitive apprenticeship model is scaffolding, where the teacher provides targeted supports that help the learner accomplish the task more independently. A key component of the cognitive apprenticeship model is the teacher's ability to provide the right scaffolds at the right time for each learner, and to know how and when to fade the scaffolding so that the learner moves gradually towards autonomous thinking. The remaining three steps of the cognitive apprenticeship model involve helping learners do the task on their own. These steps involve articulating an explicit understanding of their problem-solving processes, reflection and comparison of their processes with those of other learners, and exploration of the processes to apply to new problems independently. These final three steps help build metacognition and transfer of the knowledge to new contexts.

The cognitive apprenticeship model builds upon Bruner's belief that all learners are capable of learning if provided the appropriate tasks with the appropriate supports, and Vygotsky's notion of a ZPD—the zone in which tasks are doable when the learner is provided sufficient scaffolding. The model provides a clear trajectory for teachers to move from introducing a brand-new skill or content area to the end goal of helping learners become autonomous thinkers. In fact, the most important part of the cognitive apprenticeship model, in my opinion, is the fading of scaffolds towards autonomy. We all need supports, but we all also deserve the chance to see what we can eventually do on our own.

So, what are you thinking?

1. What are examples of the prior experiences you use to construct new knowledge?
2. What is an example of a distributed learning experience you have been involved with?
3. In what ways have you been either the apprentice or the master in a cognitive apprenticeship experience?

Learner-centered Learning

In the 1990s, around the time I joined TERC, a new field called *learning science* was emerging. Rooted in the work of Piaget, Bruner, and Vygotsky, learning science uses rigorous and systematic research methods to study theoretical foundations of learning. Earlier psychological research tended toward clinical tests with small numbers of subjects in a laboratory setting. Educational statistical research, on the other hand, was tending towards large-scale quantitative measurement through widespread standardized testing. Learning scientists advocated for balanced research agenda framed in theory of cognition and learning while also using a rigorous combination of qualitative and quantitative methods within the context of real classroom situations.

The U.S. National Research Council issued a series of books synthesizing the early learning science research, including a book called *How People Learn*. While critiqued by some for not considering a diverse enough learning audience, this work has been fundamental to much of the learning science theory that has evolved over the past decades. The big takeaway from this work is that learning is a *learner-centered* phenomenon. That may seem like an obvious statement at first. But when you consider how much of school activity is typically driven by the teacher, testing, and a fixed curriculum, it is clear that students are still not at the center of most of today's teaching and learning experiences.

A call often heard in the early learning science community was to move the teacher from being "a sage on the stage" to more of a "guide on the side." This sentiment resonates with the work of Dewey, who described the teacher as a facilitator of the learning process. In a time still shaped by normal schools, Dewey argued that learners' social conduct and individual thought should be balanced, with schools allowing time for scientific thinking and creativity, and that students could find their own purpose and agency in lessons when they have an influence on how the teachers would augment and improve the learning experience.

This emphasis on learner-centeredness is also echoed in the efforts of Maria Montessori, who designed entire learning environments with the child at the heart of the learning process. Materials, furniture, and everything else that a learner encounters in the Montessori classroom is meant to be sized for the learner and designed to be developmentally appropriate to promote learner belonging and agency in the learning process. Montessori students are free to choose their own learning experiences and learning groups, shifting among them fluidly as they desire, while teachers guide them to build skills to learn on their own. The Steiner Theory of Learning, which has since grown into the large network of Waldorf Schools around the world, also follows this interest-based model of education through a phased model of discovery learning and student autonomy, with a three-phase model emphasizing sensory learning, emotional learning and creativity, and finally independent intellect and critical thought.

When entering a typical US or Canadian classroom in a public school, especially in STEM classrooms where I spend most of my time, what I usually observe is vastly different from what these models describe. To a large extent, the teacher is standing in front of the class laying out a set of instructions for a prescribed activity. The activity may be hands-on, maybe not, but it is rarely driven by a student's questions or interest. It is rare to find students conducting the same practices as a scientist or engineer in class. Cognitive scientist Derek Cabrera notes in a 2011 TEDx talk how even playtime has become prescriptive. He observes how construction toys such as LEGO blocks increasingly come in kits with preset instructions so that kids don't make whatever they can imagine anymore. He attributes this tendency as a cause of why undergraduates are coming to university unprepared to solve problems. It certainly goes against all that we know about how people learn.

So, what are you thinking?

1. What experience have you had as a learner where you felt you were at the center of the learning process?
2. How have you provided learner-centered learning experiences to others?

Bibliography for Chapter 5

Bransford, J. D., Brown, A. L., & Cocking, R. R. (2000). *How people learn* (Vol. 11). National Academy Press.

Cabrera, D. (2011). *How thinking works* [Video]. TedxWilliamsport Conference. https://www.youtube.com/watch?v=dUqRTWCdXt4&ab_channel=TEDxTalks

Collins, A., Brown, J. S., & Holum, A. (1991). Cognitive apprenticeship: Making thinking visible. *American Educator, 15*(3), 6–11.

Dewey, J. (2011). *Democracy and education*. Simon & Brown.

Hutchins, E. (1995). *Cognition in the wild*. MIT Press.

Hutchins, E. (2000). Distributed cognition. *International Encyclopedia of the Social and Behavioral Sciences: Elsevier Science, 138*, 1–10.

Piaget, J. (1964). Cognitive development in children development and learning. *Journal of Research in Science Teaching, 2*, 176–186.

Vygotsky, L. S. (1978). *Mind in society: The development of higher psychological processes*. Harvard University Press.

6

The Role of Emotion in Learning

Ridge Road Elementary—A Grade 6 Project in the 1960s

My own public schooling took place in the late 1960s and early 1970s. In most of my elementary schooling, our classroom consisted of rows of evenly spaced individual desks, a piano in the back (where the national anthem was played every morning), and a teacher up front. The teacher's desk was a throne of authority.

But the late 1960s was also an experimental time in education. In grade 3, we spent one month in an "open classroom" format where all three grade 3 classrooms were set up with learning stations, and we were allowed to roam as we wished throughout the day. My most vivid elementary education memory is from my grade 6 class when we were charged with making toys to be sold at the annual school fair. This was a much bigger deal than the little holiday crafts we had made for our parents in the earlier years. This was a major construction and marketing project that our class worked on collaboratively for the entire month of May.

I presume the concept of making a toy for the school fair was probably our teacher's idea, but beyond that, it felt like everything was up to us students to decide. We selected a pattern to make a wooden top with a wooden handle and a string. We chose the scale, the materials, and the artistic design of the toy. We built test models and refined the design until it we got it working smoothly. We then devised a plan to make 100 consistent copies of the model. I can still remember making painstakingly detailed drawings of the toy, pressing through two pages of carbon paper so that we could create a pattern for assembly-line manufacturing.

Next, our class had to investigate the types of materials we could access in our local hardware shop. To this day, whenever I smell raw lumber, I am transported back to the woodworking store where a very patient salesperson answered all our questions. We charted out what was most affordable, given our criteria. We learned how to price out the different options and how to assess economies of scale. I am sure our teacher was there guiding us, of course, but I don't remember her as part of the process. She was an invisible support. I do remember parents joining in the effort. We didn't have computers with Microsoft Excel back then, so we used an accounting sheet that someone's dad had showed us. He taught us how to keep track of our expenses and proposed revenues.

When we had gathered all our supplies and created our business plan, we set up an assembly line in the school gym to build the toys (with adequate guidance over the use of power tools, I assume, as there were no lost fingers that I remember). Now that I think back on it, I am guessing it was not the first choice of the PE teacher to let us take over his space for the week, and I suppose it was a nightmare for the janitor to clean up after us, but somehow the school staff rallied around the project and made it all happen.

Our class sold all the handmade tops at the school fair, having determined a fair and marketable price. I remember one girl in the class had brought in the toy section of a Sears catalog so we could find comparable market values. We ensured there was a reasonable profit margin so that we had a something substantive for our efforts. Our class even worked with the PTA to choose the school projects, as well as a local charity, that would be funded by the revenues.

Today, I suppose, we would have shopped for our supplies online and likely would have found a computer-automated design tool to design and print our patterns. Maybe we would have even 3D printed the toys, or at least the prototypes. I wonder if the smell of extruded plastic from a 3D printer will trigger a future student's educational memories, as sawdust does mine. Whatever the trigger, I hope decades from now learners everywhere will have such rich and engaging educational experiences to remember.

Emotional Engagement in Learning

Psychologists would point out that the reason I remember so much about my grade 6 school fair project is because I was emotionally invested in the experience. I was stimulated by the new sights and smells of the lumberyard, and I was socially invested in working with my classmates. My creativity and intellect were engaged in a challenging problem. I took pride in my accurate drawings, and I was motivated and given a sense of purpose by giving back to my community. The learning experience tapped into many of my emotional response systems.

The role of emotion in learning is particularly deep and complex. Neurologically, emotion can be thought of as a combination of feelings, physical responses, and behaviors that result from a trigger, a stimulus that we perceive as personally significant. This trigger sends signals to other regions of the brain and provokes a cognitive reaction using the same networks and processes our ancient ancestors used to flee from a perceived attacker or to love a newborn enough to sacrifice our own safety for its survival.

When a learner's senses and emotional responses are triggered, this may capture and sustain the learner's attention, motivate them, and activate their learner's high-order cognition networks, which, ideally, all work together simultaneously and harmoniously with the executive function processes. This strong link between emotion and learning suggests that engaging students' interest and emotional responses should be paramount in the early design of learning activities. In fact, educational neuroscientists such as Mary Helen Immordino-Yang suggest that the fundamental purpose of education may be to help build students' capacity for regulating their emotional responses and behaviors using increasingly flexible strategies to deal with a variety of complex situations. They argue that these are the very strategies that promote creativity in art, science, and innovation, which are unique to our species.

Motivation

Emotional connections are critical for *motivation*, which is the impetus that gives purpose to our actions. Motivation stems from a desire or need to take action, along with the beliefs, perceptions, and values that accompany that desire. Motivation can be extrinsic—such as an award, money, or approval—a reward that comes from an external source, or it can be intrinsic, which is driven by the pleasure of the activity itself. Either type of motivation can foster engagement—the state of being involved or immersed within an activity, but there is evidence that intrinsic motivation is more conducive to deep engagement and learning.

Fostering an emotional connection doesn't mean that everything should come easily. Learning and engagement also rely upon some challenge. Balancing frustration, boredom, and anxiety with challenge, support, and risk are all part of design of learning experiences to keep students motivated and in their ZPD. When we are in this optimal zone, sometimes we can get lost for hours. Nearly all the neurodivergent STEM professionals whom I interviewed for this book said they get lost in the work they do. They talked about losing track of time and forgetting about other things when they were absorbed in a task. They said that's when they feel most productive.

Flow

The psychologist Mihaly Csikszentmihalyi described the state of total engagement as *flow*—a focused state of concentration that can lead to optimal productivity. Csikszentmihalyi and his team ran studies with thousands of adults and children to understand the contributing factors to flow and learning. Researchers observed classroom activity, interrupting the activity briefly with a timed bell so they could have spontaneous check-ins on what the teacher was discussing or doing with the class, and what the students were thinking about and feeling at the time. During an overwhelming majority of the time, students were attending to something completely different from the lesson (e.g., food, after-school activities, the person next to them). The observers also found that the students rarely exhibited any behaviors indicative of flow.

Csikszentmihalyi suggested that intrinsic motivation is crucial to maintaining a flow state. Tasks conducive to flow also typically have clear and apparent goals. This is similar to why people find themselves looking up after hours at a jigsaw puzzle or sudoku—tasks where each next step is right in front of them—and realize the sun has set without them noticing. Similar to Vygotsky's notion of a ZPD, Csikszentmihalyi noted that flow requires that a task be in a "sweet spot" of feasibility, where it is within the potential capability of the learner yet still hard enough to present a challenge. Designing learning experiences is a bit like a Goldilocks problem, where we are striving for just the right balance between too difficult and too easy. If a task is too simple, it becomes tedious and boring. If it is too hard, it may be daunting and even overwhelming. Finding that sweet spot is critical for deep and meaningful engagement.

The trick, however, is that the sweet spot to engage and motivate learning may be different for every learner, and for some learners it may even change from day to day. Learning experiences and environments must be flexible enough to hit multiple sweet spots for multiple learners. This is not easy and calls for differentiated teaching and learning, which will be explored in Chapter 9.

So, what are you thinking?

1. What motivates you to learn something new?
2. When was the last time (or any time) you have felt in flow?
3. What types of learning activities do you think engage your students in optimal learning? Is it the same for all your students?

Becoming a STEM Problem Solver

Another reason the toy-making project during my grade 6 class was so successful is because of the sense of empowerment I had in the process. I felt like I was an integral part of the team that made it happen. I contributed to decision making, and I experienced the direct impact of those decisions in the process. I took on the roles that suited me, and I was given the space to do things my way and learn from my mistakes. The experience offered an important opportunity for me to build identity, agency, and autonomy as a problem solver.

STEM Identity and Learner Agency

Learning involves not only construction of knowledge, but also the construction of a sense of self, which is referred to as *learner identity*. Learner identity can be thought of as how a learner feels about themselves as a learner. Related to identity, *learner agency* is the extent to which the learner feels in control of their own learning and is inherent in their ability to regulate, control, and monitor their own learning. Learners employ agency to manage their cognitive, affective, and behavioral processes as they engage in learning activities.

STEM education professor Angela Calabrese-Barton and her colleagues explain that STEM learner agency intersects with race and cultural differences, and can be developed through the exploration of culturally meaningful scientific investigations. *STEM identity* is the concept of fitting in within STEM fields, relating to how individuals make meaning of science experiences and how society structures possible meanings. People with strong STEM identities may think about themselves as STEM learners and develop an identity as someone who knows about, uses, and sometimes contributes to STEM. Agency and identity are complicated in fields such as STEM that have been historically dominated with white men.

Learner agency and identity are particularly important in STEM, where learners often become anxious and alienated. For example, research in mathematics education shows that many of the teacher-led instruction styles used in much of today's math curricula tend to favor passivity over agency. Teachers often present a "correct" method or set of methods to be used, thus discouraging learners from exploring their own alternative pathways to solutions. When a teacher treats a student's contribution as unacceptable or incorrect, the student may stop trying, feeling they are incompetent. This can invoke what's known as math anxiety or math phobia—hyper-negative emotional activity that is counter to agency in mathematics—which may prohibit meaningful engagement in math learning.

Learner agency is also inextricably tied to a learner's executive function—their effectiveness to regulate, control, and monitor their own learning. A learner who is highly invested in an activity and identifies strongly with that activity may persist despite executive function challenges. There is some thought that the very act of this persistence adds to the strength of some neurodivergent learners to deal with challenging situations. Overcoming working memory gaps or distractions in an activity that matters deeply to a learner may help them recognize and build strategies that help them, even when they aren't as intrinsically motivated.

Allowing learners to pursue their own interests and make choices in their learning experience is key to helping learners develop agency. This can be as simple as letting learners choose the type of book they would like to read or letting them create a video rather than a paper. Agency is further developed by encouraging learners to set their own goals and targets for learning, particularly to gain insight into their own strengths and weaknesses related to learning so that they can monitor and adjust their practices to optimize their own unique learning process. Learner agency also is supported by creating an environment where learners are able to communicate effectively with teachers or peers, especially to ask for help when they need it. Learners need to feel it is a useful part of the learning process to ask for help, not a threat or exercise in futility.

Learner Autonomy

When a learner is in the driver's seat, particularly when they are supported with the appropriate scaffolding (and those scaffolds are appropriately faded), this ultimately can lead to *learner autonomy*—the ability to take charge of one's learning. Learner autonomy is key for intrinsic motivation in the learning process.

For employees in STEM neurodiversity programs, their desire for autonomy was expressed in terms of wanting a workplace that allows them to "go away and think about a problem for a while" and "to do it my way." Recognizing that their own methods of approaching a problem may differ from the rest of the team, many of the neurodivergent learners I spoke with needed time and space to think alone first. They didn't like having to spend time defending, or even explaining, their methods at the same time they were still working through the problem. They said the task of having to align their ideas to the thinking of others was an extra burden, taking energy and attention away from the actual solving of the problem. Some others felt that their different ways of thinking might be criticized before a solution was clear. When they were given the time to make their argument first, they found their ideas were more readily accepted by the team.

Learner agency and autonomy still require a strong element of planning and communication. It requires *metacognition*—being aware of one's own learning. Metacognition describes learner's awareness and attention to the larger goal of a task and an ability to monitor progress towards that goal. Metacognition can be particularly difficult for learners with certain limitations in executive function. Educators (and employers) can play an important role in helping learners develop strategies for metacognition. Tools such as calendars and graphic organizers can help learners to keep on task, stay focused, and support their own working memory and attention towards the goal. These tools can be used as supports to help learners focus on the aspects of problem solving where they shine.

So, what are you thinking?

1. Who has agency in your classroom or workplace? What does that look like?
2. What elements of your life are most important in your own identity and agency?
3. When you are given autonomy of thought, how does that feel different from when you are following someone else's procedure?

Anita—Embracing Rabbit Holes

When I first met Anita, she introduced herself as a neurodivergent entrepreneur. Anita said she struggled in school but got good enough grades to get by. She remembers being struck on the hand by a teacher when she wouldn't sit still in piano lessons. She had always longed to be out playing in the sunshine or in the rain. It didn't matter as long as she was outside and as long as she could move. As an adult, Anita was diagnosed with ADD, something that never had occurred to her parents or her teachers to investigate earlier because she was smart.

As a child, Anita had a world of imaginary friends who kept her entertained. She would build forts in the woods and even cook meals for these friends over small fires she'd start by herself with a matchbook she'd stolen from her brother. That was until the fire department had to be called because a small patch of the woods, including her fort, was burnt to the ground. She remembers being in trouble for a lot for things like that. Her mother called her a "bull in a China shop" because she was always breaking things.

Anita also remembers building a snack stand in her front yard so she could earn money for the local children's hospital. Each year she'd set up the stand for three days in the summer—once in June, once in July, and once in August—and she'd sell lemonade or fruit punch, cookies or potato chips to the passersby on the busy street where her family lived. She liked earning the money, but she really liked the responses she got from the neighbors. She liked when they called her "a good businesswoman" and "ahead of her years." She felt good knowing she was doing a good thing by giving the money to the children's hospital. She also remembers thinking, even at a young age, that the lemonade stand was more fun if she did it only once each month in the summer and if she changed up the menu each time she did it. She didn't want it to be the same all the time. She didn't like things to get boring.

In high school, Anita started a sports exchange program with her friends, where they swapped sporting gear as they grew and changed interests. "I had three pair of unused cleats in my closet when I stopped playing soccer, and I really wanted to try skiing," Anita explained. "The Sports Swap was an idea that came to me as a way I could get my hands on a used pair of skis, but it took off from there. It became way bigger than I could handle." As the popularity of the Sports Swap grew, Anita handed it off to a few other kids to run. They made it part of the school's Spring Fair. Anita went on to think about new things she could do for the community.

After high school, Anita took courses in animation and media production at the community college while she also participated in parkour in a downtown club. Her time was split between playing and building games in virtual worlds and extensive outdoor exercise, usually at least four hours per day. The COVID pandemic found Anita living in a 400-square-foot apartment in a high-rise she couldn't escape. During this time, quarantine lockdowns in her city sometimes even prohibited a walk around the block after her workday was done, which was extremely difficult for her. She knew she needed a change. She also realized that the work she was doing for an advertising company could be done from anywhere, and there was a growing market for her skills in virtual reality (VR).

Anita had been playing with an online group of gamers who invited her to join them in a house they'd found in a small city where rent was cheap, COVID numbers were low, and tech start-ups were booming. I asked her if she had been fearful of leaving the life she knew and starting over in a new place. Anita laughed and said: "I figured if not now, when? You've got to live while you can." She was ready for an adventure.

Anita packed up her things and moved, expecting to land a job similar to the one she'd already had. Instead, within six months, Anita ended up starting her own VR company and growing it one client at a time into a business that now has a staff of six. Anita says she never could have done it without the help of her roommates, who basically filled out all the forms for government support and loans. Anita knew what she wanted to do, but she says she didn't have the patience to sit down and get the paperwork done. The first person she hired was a full-time administrator.

Anita says she still struggles with what she calls rabbit holes. She gets an idea and sometimes she will stay up all night poring over the idea, researching it and sketching it out in different ways, even past her own sense of productivity. She says sometimes she knows it's a bad idea, but she just feels compelled to see it through. She has to get to a point where she can sketch out the whole idea, to the point where she can explicitly see its flaws, and then she can put it to rest. "I know some would see that as a waste of time," she says, "but I won't be able to move on until I do it." She also says some of her best ideas come to her spontaneously when she is down one of those rabbit holes.

Anita and her team are now creating artistic and human-centered VR experiences for companies to use for remote training. She'd like to create more VR experiences that are designed just for pleasure and exploration that she can install in nursing homes, hospitals, and other places where people are confined. She'd like to use her skills to improve the lives of others, and most importantly, she knows she always wants to be doing something new.

Bibliography for Chapter 6

Boaler, J., & Greeno, J. G. (2000). Identity, agency, and knowing in mathematics worlds. *Multiple Perspectives on Mathematics Teaching and Learning*, *1*, 171–200.

Bransford, J. D., Brown, A. L., & Cocking, R. R. (2000). *How people learn*. National Academy Press.

Calabrese Barton, A., & Tan, E. (2010). We be Burnin'! Agency, identity, and science learning. *The Journal of the Learning Sciences*, *19*(2), 187–229.

Carlone, H. B., & Johnson, A. (2007). Understanding the science experiences of successful women of color: Science identity as an analytic lens. *Journal of Research in Science Teaching*, *44*(8), 1187–1218.

Code, J. (2020). Agency for learning: Intention, motivation, self-efficacy and self-regulation. *Frontiers in Education*, *5*, 19.

Csikszentmihalyi, M. (1997a). Flow and education. *NAMTA Journal*, *22*(2), 2–35.

Csikszentmihalyi, M. (1997b). *Flow and the psychology of discovery and invention*. Harper Perennial.
Holec, H. (1981). *Autonomy and foreign language learning*. Pergamon Press.
Hughes, R. M., Nzekwe, B., & Molyneaux, K. J. (2013). The single sex debate for girls in science: A comparison between two informal science programs on middle school students' STEM identity formation. *Research in Science Education*, *43*(5), 1979–2007.
Immordino-Yang, M. H., & Damasio, A. (2007). We feel, therefore we learn: The relevance of affective and social neuroscience to education. *Mind, Brain, and Education*, *1*(1), 3–10.
Tyng, C. M., Amin, H. U., Saad, M. N., & Malik, A. S. (2017). The influences of emotion on learning and memory. *Frontiers in Psychology*, *8*, 1454.

7

Extraordinary Learners

Addison and Krystal—Bursting With Excitement

Two girls at Hillside Junior High—I'll call them Addison and Krystal—were both extremely bright, very fun to be around, very helpful at times, and at other times, extremely disruptive in school.

Addison was a petite girl, with tiny hands and tiny feet, and a very loud voice. Many days, just as Ms Bradbury's class settled into an activity, the door would fly open and Addison would come flying in. She would usually be yelling something like "Miss! Miss! Ray took my phone and now I can't call my mother and she *needs* me to call her by 10 am. Miss! Miss! Ray took my f*in' phone!" A sympathetic reader might worry that there really was something urgent that Addison had to call her mother about—a fair concern—except that this behavior happened nearly every day in every class, and Addison's outbursts could have just as easily been about a nabbed candy bar or anything else. Addison reacted to many daily life events as if the world were ending.

Addison was also extraordinarily bright. She read sophisticated novels, far beyond the class reading level, and had long interpretive conversations with her English teacher. She was good at building on things that she already knew. She caught on to new concepts quickly and saw connections between concepts that led her to far more complex ideas than many of her peers. She was a beautiful writer and visual artist. Creativity oozed from her pores, on her good days. Even while missing inordinate amounts of class time due to medical appointments and disciplinary meetings, Addison was able to keep up with her schoolwork—at least for the first year at Hillside Junior High.

Over the next two years, however, Addison was increasingly pulled out of class because of behavioral issues. When asked to sit still and wait for others to gather their thoughts, it was like her nerves got the best of her. More times than not, she blurted something out, cutting off someone else, things escalated, and someone said something that caused her to explode. While she produced outstanding results on work in class when she was attending class, with all her behavioral interruptions, Addison eventually missed so much work her grades started to suffer.

Addison was invited (or told) to sit in the learning center to calm herself when she got upset in class. After a while, she ended up spending more and more time in the learning center, sometimes by her own request. In another half of the learning center was the space where learning specialists helped the kids who needed reading and math support. Addison was allowed to move around in the learning center and she started helping to support other kids. She often was more knowledgeable about the class content (such as assigned math problems) than the learning specialist and Addison seemed to enjoy teaching others. In the learning center, she was also able to move freely from one group to the next, blurting out whatever seemed to help at the time. She was able to be herself.

Unfortunately, however, being pulled from class as frequently as she was became a vicious spiral for Addison. She sat alone in the rare time when she was in the classroom, and she seemed to be losing her connection with peers and classroom teachers. Addison was also accessing the public health system and coming to school with constantly changing medications. There were new meds, changing dosages, and sometimes no meds because they were forgotten at home. There were days when Addison literally did not stop talking from the moment she entered the school to the moment she left, and there were other days when she slept through class, head down and out cold. There was very little communication between Addison's family or health care workers and the school, so the changes in Addison's behavior due to medication were unpredictable to her teachers. Similarly, the doctors likely knew very little about Addison's behavior (and strengths and talents) at school. That dearth of communication left Addison falling through the cracks of the system meant to help her.

I lost track of Addison after Hillside and wasn't able to follow up to know what has happened to her after high school. She dropped contact with most people we knew in common, which usually isn't a good sign. I like to think Addison found her way wherever she landed, but I wouldn't place too high a bet on it.

Krystal was the physical opposite of Addison. Krystal was about 5'10" tall, solidly built, and was the only child at Hillside that ever scared me. There were days I wouldn't be alone in a room with Krystal. On the other hand, there were many days when she and I could be hugging and laughing like old friends. She could truly charm me like no one else.

Krystal often settled right into class. Sometimes she was even the first one with her notebook open and ready to learn. She prided herself in her keen intelligence, and she was good at school. That is, until somebody did something to tick her off. Krystal could go from working diligently on an assignment to an all-out kicking and screaming fight in a flash. One person just had to push the right button, and she exploded. In this tight-knit group of kids, many of whom had grown up together in an overcrowded public housing community, everybody's buttons were well known and right out there for the pressing. Krystal was sometimes held in the office, sent home, or even referred to the police in an effort to keep the other children safe.

Hillside teachers started to notice Krystal's struggles increased around mid-October when the weather turned cold. Before then, Krystal could often be found shooting hoops on the basketball court early in the morning before school and again over the lunch hour. Sometimes she snuck out there between classes for a few quick throws and a runaround. When a storm hit in October with several days of really foul weather, Krystal started to act out. That happened to be the week I first met her. She was strong and angry, and she had a very short fuse.

As soon as basketball season started in November with daily practices in the gym, Krystal settled down a bit and her talents reappeared in class. There were still outbursts from time to time, but Krystal had in-class supports. Her math teacher was her basketball coach and her social studies teacher also taught her PE teacher, so they had gained Krystal's trust, spurred on by their love for athletics. Krystal would go talk with them when she started getting upset rather than waiting until it was too late. They both gave her extra time to work her through her issues at lunch and after class. They built on their existing rapport through basketball to coach her emotionally and socially as well. Together they became Krystal's support team so that she was rarely removed from class because of behavior.

In math class, Krystal often sat at the back table where the kids who needed help worked with a volunteer. Krystal was able to fly through the math problems, but she liked the personal attention at the help table. She said it helped her keep focus. It also helped her stay out of trouble. She, like Addison, became a resource for other students at the table. She would explain how she solved the problem in language that the other kids could understand. This shift in her role from a troublemaker to a tutor gave Krystal a chance to see herself as bright. She was doing it in the context of her classroom, however, not being pulled out to another room like Addison had been. She became valued as a contributing member of the class by her peers and her teachers. She was not just a basketball player who tended to start fights; she was a bright math student who could help others.

Krystal went on to do well in high school. She was a key player on the basketball team and received a scholarship to a university. She came back to visit Hillside when she was home to visit her old teachers. She credits them with keeping her on the right path. She is studying business and aspires to start a business related to sports.

Krystal and Addison's stories have so many serendipitous turns of events that one can never know what exactly predicts success or derailment. But what we can observe is that when the environment works against the learner, it often doesn't go well. When a child is constantly pulled from class and isolated from the learning experience, and constantly being told it is their problem that they can't learn, it is an uphill battle for that child. On the other hand, when a teacher has the time and the connection to work with the learner, supporting their physical, social, and emotional needs and helping build a sense of belonging within the classroom, not only may their immediate engagement and productivity improve, they may also learn to see themselves as good learners in the long run.

Twice-exceptional Learners

By the time I taught at Uni High School in the 1980s, the small school had already graduated three future Nobel laureates and many renowned authors, activists, and innovators. Many Uni students were extraordinarily talented in music and/or art and/or math and/or something else. Many of the same students seemed to struggle excessively with keeping their papers organized or even remembering where their locker was. I taught a brilliant future physicist who couldn't sit still in a chair without falling asleep. I taught a student who intricately and accurately explained the origin of the universe through an interpretative dance, but then bombed out on the final written exam. One of my students wrote an extraordinary paper on the Tao of Physics that opened my eyes to new meanings, even though I had already read Fritjof Capra's popular book more than once—but then he never completed another assignment in my class.

Looking back, I realize that many of my students were likely what is now called twice exceptional or 2e—students who are immensely talented in some areas, and yet also struggle with other aspects of traditional formal education. Variation in brain function—be it due to genetic or environmental issues including health, stress, or trauma—can manifest in behaviors that seem to be at both of these cognitive extremes.

Overexcitabilities
Polish psychologist Kazimierz Dąbrowski described a set of *overexcitabilities* that guide much of the research on gifted or

twice-exceptional learners. These overexcitabilities include psychomotor, sensual, imaginational, intellectual, and emotional extremes. Dąbrowski argued that when these overexcitabilities are combined with innate ability and intelligence, they are predictive of potential for higher-level achievement. Further research has indicated that overexcitabilities—especially in imagination, intellect, and emotion—may cause a person to experience daily life more intensely and to feel the extremes of the joys and sorrows of life profoundly such as having intense sensual experiences, creativity, and stamina and motivation for this creative work. In the same vein, overexcitable individuals may also experience increased stress because of their heightened sensitivity to the world around them.

The overexcitabilities most directly connected to STEM problem solving are intellectual and imaginational overexcitabilities. *Intellectual overexcitability* can stimulate a learner's desire to seek understanding and truth, to gain knowledge, and to analyze and synthesize. Learners with intellectual overexcitability are often avid readers, questioners, and observers. They may conduct elaborate plans and have detailed visual recall, while also being able to think abstractly and question different thoughts of morality and fairness in local and global issues. Intellectually overexcited learners often thrive when they are allowed to question and analyze, and when they are given a way to act upon their concerns.

Imaginational overexcitability is characterized by frequent and rich uses of image and metaphor, invention and fantasy, detailed visualization, and elaborate dreams. Learners with imaginational overexcitability may "live in an imaginary world" and seem not to separate fact from fiction. Problem solvers with highly active imaginations may see innovative and creative connections between ideas and may tend to be able to abstract their knowledge to see new solutions to new problems.

Disruptive behavior may occur when intellectually overexcitable learners' train of thought runs ahead of those around them and they then become impatient while waiting for others to keep up. This can cause impulsive interruptions or loss of interest. Providing autonomy of thought and expression—by allowing students time and space to work through their own conceptual knowledge through the representations, scenarios, and media of their choosing—may help imaginational overexcited learners thrive in the classroom. At the same time, intellectually overexcited learners also may need supports to learn how to make room for others' viewpoints without rushing ahead to a conclusion. This is yet another way teaching and learning activities can be differentiated for individual learners.

So, what are you thinking?

1. Do you or someone you know exhibit overexcitabilities? How do they manifest themselves?
2. What extraordinary talents do you or your students bring to STEM problem solving that may be masked by other challenges?

Creativity

Many of the strengths and talents described by employers and employees in STEM neurodiversity programs are related to important habits of mind for problem solving outlined by STEM education professor emeritus Art Costa. One of these habits of mind is creativity. The idea of creativity came up in nearly every conversation I had about neurodiversity in STEM. Many STEM employers said that neurodivergent employees' creativity is one of the talents they considered most useful. They also said that having diverse perspectives, in general, added to the overall creativity of a team.

The American Psychological Association defines creativity as the ability to produce or develop original work, theories, techniques, or thoughts. Creativity is also related to both intellectual and imaginational overexcitabilities. In neuroscience research, creative cognition is shown to use both the default network, functionally associated with self-generated thought and spontaneous thinking and the executive network, functionally related to executive function.

Engaged and effective problems-solvers often solve problems because they are genuinely interested and compelled to find a solution. Costa adds that creative problem solvers often delight in making up problems to solve on their own and request enigmas from others. They enjoy figuring things out by themselves and continue to learn throughout their lives. Effective problem solvers often use analogies and past experiences to help them puzzle through new situations. They can abstract meaning from one experience and transfer it in a new and novel situation by seeing connections and patterns among problems and abstracting across problems to find generalizable solutions. They are also often uneasy with the status quo, so they seek novelty and improvement. Creative problem solvers take educated and reasonable risks with their creativity while remaining open to criticism.

A standard test of creativity, which measures a person's divergent thinking—the act of generating multiple ideas from a single starting point—is called the Guilford test. In the Guilford test, the person is presented with an everyday object such as a paperclip, a coffee mug, or a spoon, and is given a limited amount of time to brainstorm as many uses for that object as possible. Guilford tests are scored in terms of fluency—how many uses the person came up with, flexibility—how many different categories or areas are covered, originality—the unusual nature of certain answers, and elaboration—how detailed and developed the answer is. Some argue that there may be other dimensions of creativity not measured through divergent thinking tests, such as spontaneous thought, so better instruments are needed to advance this research.

Researchers have found that pre-existing knowledge can often be an obstacle to creativity. People can get stuck in old ways of thinking and get fixated on a specific constraint or idea. For example, when designers are shown examples before being asked to invent something new, their inventions tend to incorporate aspects of the examples. Some neurodivergent learners, however, may be less constrained by previous knowledge as the general population, making them exceptionally good at creative tasks, such as inventing creative new uses and brainstorming new features for everyday objects. Divergent thinking is thought to be a particular strength for some people with ADHD. For example, when shown a set of example toys that shared specific features (e.g., a ball) and then asked to invent new toys that were very different from any existing toys, people within the ADHD group were less likely to replicate similar features as the examples provided.

Psychologists Torrance and Dauw first noted an association between highly creative children and behavioral issues. It was thought that the behavioral issues might be a product of their repressed creative needs. Later researchers hypothesized that the low arousability associated with ADHD may lead to higher drives for sensation-seeking and more stimulation-seeking behaviors, which in turn may lead to greater flexibility, openness to experience, preference for complexity, and receptivity to novel ideas and experiences. In the classroom, creative learners with imaginational overexcitability may easily become bored and may doodle or create stories to escape from the tedium of schoolwork. The school environment needs to change to embrace this creativity rather than stifle it.

So, what are you thinking?

1. In what kinds of activities and under what conditions do you feel most creative?
2. What conditions tend to stifle your creativity?

Systems and Patterns

While the broad and expansive thinking involved in creativity is a strength for some neurodivergent learners, others may excel in fine-grained, detailed pattern recognition. Patterns were used by animals and early human brains to identify food and distinguish them from sources of danger and therefore our brains evolved to make these distinctions efficiently and effectively.

The measurement of pattern recognition is often conducted with two types of cognitive tasks. The first is the Embedded

Find the shape:

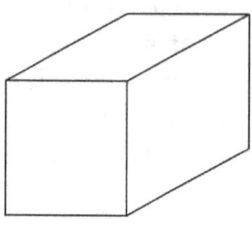 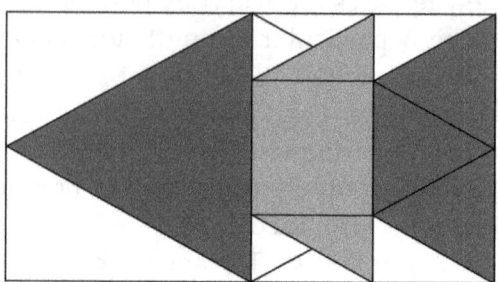

FIGURE 7.1 An Embedded Figures Test where the task is to find the embedded rectangular prism within the diagram

Figures Test (EFT), where the person is to find a specified visual–spatial pattern within a broader scene.

In this test, you are asked to find the simple pattern (the rectangular prism) within the complex pattern. It requires your brain to mask out all the irrelevant complexity and attend solely to the lines that make up the simpler pattern.

A second type of test commonly used is a Raven's Progressive Matrix (RPM), which is often found on traditional intelligence tests. The RPM requires prediction about what pattern comes next (or is missing) from a series of systematically changing tiles (see Figure 7.2).

The RPM measures fluid intelligence, which is the ability to think logically, identify patterns, and solve problems independently of acquired knowledge. Both the EFT and RPM are similar to items found on typical intelligence tests, and performance on either test has been positively correlated with general intelligence. In STEM problem solving, recognizing patterns within groups of problem solutions enables the abstraction of solutions that can lead to algorithm design that is at the heart of computational thinking. Detailed pattern recognition is necessary in cybersecurity, where data analysts must sift through vast amounts of information, focusing on salient trends and patterns.

There is evidence that autistic people score higher on the EFT and the RPM than the general population. Pattern recognition has also been proposed as a primary basis for particular talents and savant behaviors in autism, such as calendar calculating,

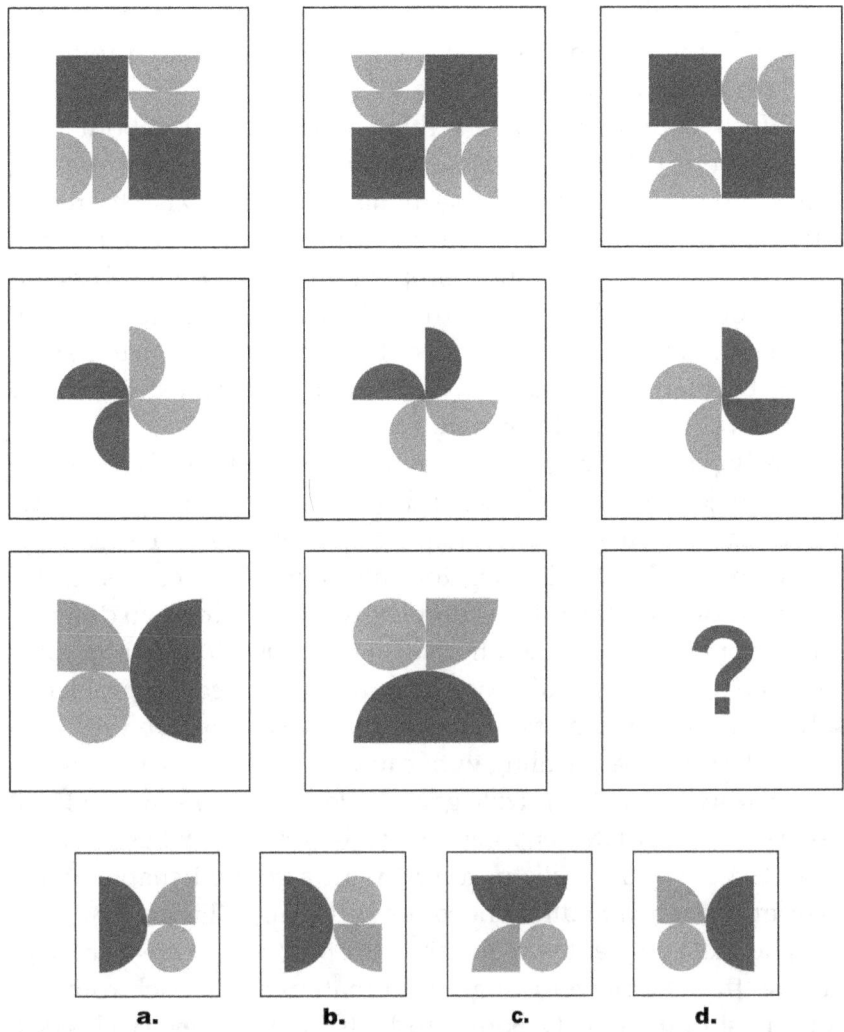

FIGURE 7.2 A Raven's Progressive Matrix where the task is to identify the missing piece in the pattern matrix

mathematics, and other specialized skills. Researchers have found a potential connection between autism and the "systematizing" behaviors required for problem solving and innovation. Baron-Cohen suggests that autistic people may be particularly good at examining and identifying causal relationships among events. He argues that the drive in many autistic people to observe carefully, analyze, and construct rule-based systems is related to their strengths in systematizing, and may provide a strength in this area.

Effective problem solvers think systematically about how to test every different possible solution to make sure nothing is left undone. This type of systematic thinking and thorough attention is related to rule-based thinking and a compulsion towards thoroughness and completion—traits often observed in neurodivergent learners. An autism researcher formerly at Microsoft, Andrew Begel, says autistic people can often see a whole problem in a visual way and often have an extraordinary ability to sift through information.

Early pattern recognition activities for children often involve sorting objects by size, shape, and color. When sorting takes into account a quantifiable dimension (e.g., size or age), then predictive arrangements into series beg the question "what comes next?" This type of task expands in complexity when using two different sets of objects (e.g., fitting differently sized lids to differently sized containers). In these rule-based predictions, effective problems-solvers invoke systematic thinking practices to ensure they consider all cases when dealing with causal relationships. In my own team's research with a puzzle game called *Zoombinis*, which I will discuss more in later chapters, there is a puzzle where players must target specific cells in a grid with a colored shape, where rows are assigned to different colors and shapes (Figure 7.3).

The most efficient way to solve this puzzle is to create a diagonal so that the player can get information about each row and column simultaneously. Some students in almost every class we studied, however, seemed to get more satisfaction out of completing the entire grid. While we did not conduct a rigorous research study of these pattern completers, their behavior is consistent with young learners thinking systematically. When I watch the joy these pattern completers get from checking every last box, I have often thought that if I were a manager of a quality assurance program, these are exactly the type of employees I would want—testers who seemed physically compelled to try every single use case and not be able to sit still until the task was complete. I suppose that's what Thorkil Sonne, who started

FIGURE 7.3 A screenshot of Mudball Wall Puzzle in *Zoombinis* where players may want to complete the pattern rather than solve the puzzle

Specialisterne, and the other STEM companies discussed in Chapter 3, were likely thinking when they sought autistic software testers. Attention to detail and systematic thinking are critical STEM problem-solving talents.

There is a growing body of research that people with dyslexia may have particular strengths in different kinds of systems thinking, particularly in spatial reasoning and seeing connections between elements of a system. There has been additional research showing that people with dyslexia may also have an increased capacity for understanding spatial relationships between objects, such as how they are positioned in space, how they move, and how they interact with one another. There is some evidence that people with dyslexia may be quicker at making connections among ideas and that they tend to see the world as made of systems, rather than their individual components. These talents are all critical in looking at the world in new and different ways, and seeing connections between actions and their intended and unintended consequences.

The future of STEM problem solving requires thinkers who pay great attention to detail, as well as those who keep a broad perspective to contextualize information within a body of existing ideas. Our STEM workforce needs teams of pattern recognizers and systematic thinkers along with idea generators and big-picture thinkers, working together to push the field in new directions. These talented learners are out there, they just need to be recognized and nurtured.

So, what are you thinking?

1. Are you or any of your students "pattern completers"? Can you describe the satisfaction it invokes?
2. In what types of learning activities might pattern recognition and systematic thinking be highlighted?

Productive Persistence

Another STEM problem-solving habit of mind, and overall life skill associated with neurodiversity, is persistence—often referred to as "stick-to-itiveness." Persistence is intertwined with attention and self-regulation. Approaching large and complex problems requires self-regulation to ensure all angles have been considered and all assumptions have been checked, as well as an ability to maintain focused attention on the task for the duration, overcoming potential distraction and possible boredom or frustration. An effective problem solver needs to keep an eye on the long-term goal and not get derailed.

Strong problem solvers often have a sense of deliberativeness that requires this self-regulation. They take time to develop a strategy for approaching a problem before starting, waiting until they understand a fuller set of information before reacting or making decisions. Taking adequate time to reflect on information before acting may, in the end, be more efficient by avoiding dead ends. This reflective pace may be very much in the comfort zone for people who can become anxious around transitions, but others who are itching to move forward on a plan may find the planning stages tedious and stressful.

Psychologist Angela Duckworth writes about the notion of grit as a blend of passion and persistence that is associated with high achievement. Persistence, in and of itself, is necessary but not sufficient for effective problem solving. A further refinement of this interpretation of persistence includes the idea of productive persistence, which differentiates between positive persistence towards a goal versus "wheel-spinning." Wheel-spinning may look like persistence, but it gets nowhere, as in repeatedly trying the same thing, even though it doesn't work. Productive persistence, on the other hand, is continuous and effective effort towards a goal.

Knowing when enough work is done to ensure accuracy without persevering beyond the point of diminishing returns is a skill for efficient and effective problem solving. This requires metacognition, the ability to know what we know and what we don't know. Costa explains that effective problem solvers know how to ask questions to fill in the gaps between what they know and what they don't know about a problem. Seeing large problems as a whole is an area where some people have particular talents while others can get buried in fine detail, losing sight of the whole.

So, what are you thinking?

1. Under what conditions are you or your learners most productively persistent?

Ms Bradbury's Grade 7 Class—Drumming Through Transitions

The transition to junior high school from elementary is often a challenging one. But in September 2021, Ms Bradbury found her new class of grade 7 students were struggling even more than usual. She said that after the COVID lockdowns, it was like these children forgot how to be with other people. The year of upheaval, and the months without structure and socialization, left many children without the resources to self-regulate. They were trampling each other in doorways, talking over one another, and just carrying on with an unprecedented level of rudeness. Ms Bradbury said in all her years of teaching she'd never seen such chaos as during the return to school after COVID.

She and I talked about executive function a lot during those months. We talked about what children need to self-regulate and what situations provoked too many barriers for them to overcome. She said transitions were the hardest. When the children came in from recess or after lunch, it took an inordinate amount of time for them to settle the children down. At the time of our conversation, I had just been reading about how, for many learners with executive function struggles, stopping and starting is hard. Parenting education specialist Shauna Tominey had just given a presentation at TERC where she introduced games such as freeze-tag to help children practice self-regulation by having to run fast, stop suddenly, and start running again. That changing of state is hard for some children, and when practiced young in games such as tag, they may be able to exercise those neural pathways for later. In older learners, using words like "pause" and "continue" instead of "start" and "stop" helped some move from one activity to another, letting them take a respite but knowing they would be able to return later.

So when Ms Bradbury said she needed help settling kids down after lunch or recess, we brainstormed ways we could help students for whom the transition from active outdoor activities to the indoor classroom had become difficult. We wanted to build them a metaphorical ramp to go from one state to another. I first suggested a fun three-minute movie or something that would stimulate their senses but still help them transition to sitting in their seats when they came in. Ms Bradbury took this idea and came up with something better. She found a video clip of a drumming routine, something the kids could do with their fingers on the desk. The drum sequence was complex enough that no one would be able to do it the first time. It took practice. The pattern also had multiple "roles" or ways different kids could contribute. Some students could count out a simple four-beat rhythm and provide a background structure to the drumming. Others could take more melodic roles and even solos while staying within the same overall patterns of the piece.

The next day, when the class came in after lunch, Ms Bradbury showed them the drumming video and explained that each day when they came inside they should start class by practicing the drumming. She said when they felt ready, they could invite the grade 8 and 9 classes in to see their performance. The class worked together, eager to come inside after lunch and start practicing their drumming. Each day, she cut it off after 5 minutes, even when they wanted to continue. She kept the drumming as a transition activity that could be used whenever it was needed.

For the next few weeks, not only did her students begin to settle into class without incident, but she also saw some of them practicing drumming together in the halls and outside. It became a matter of class pride. She watched them collaborating to get better. Some of them started improvising new beats. The activity had engaged students and given them something to come inside for after lunch, and something to be proud of. It was a transition they chose to make.

By the winter holiday break a few months later, Ms Bradbury and I got together to reflect on the semester. She was enthusing to me about how wonderful the grade 7 students were, how brilliant they were as science learners, and how innovative they were in tech ed. She also talked about a growing collaborative dynamic in the class, how they worked together and supported each other without the competition and alliances she'd found in previous years.

During those reflective conversations, Ms Bradbury hardly remembered the outlandish brutes she had told me about in the beginning of the school year. In fact, she was startled and laughed when I reminded her. It was all so distant to her at the time. We thought about the drumming activity. We couldn't say if that was the cause of the students' successful reconnection, but neither of us could imagine how the resettling would have occurred without the drumming. It was as if these kids who had gone through the COVID lockdowns and turmoil together just needed some transitionary structure, and they were happy to be drumming to their own beats, all together again.

Bibliography for Chapter 7

Abraham, A., Windmann, S., Siefen, R., Daum, I., & Güntürkün, O. (2006). Creative thinking in adolescents with attention deficit hyperactivity disorder (ADHD). *Child Neuropsychology, 12*(2), 111–123.

Baron-Cohen, S. (2020). *The pattern seekers: How autism drives human invention*. Basic Books.

Costa, A. L., & Kallick, B. (2000a). Discovering and exploring habits of mind. *Explorations in Teacher Education, 36*, 36–38.

Costa, A. L., & Kallick, B. (2000b). *Integrating & sustaining habits of mind: A developmental series, book 4*. Association for Supervision and Curriculum Development.

Crespi, B. (2021). Pattern unifies autism. *Frontiers in Psychiatry, 12*, 59.

Dąbrowski, K., & Piechowski, M. M. (1977). *Theory of levels of emotional development*. Dabor Science Publications.

Duckworth, A. (2016). *Grit: The power of passion and perseverance*. Scribner.

Edwards, A. R., & Beattie, R. L. (2016). Promoting student learning and productive persistence in developmental mathematics: Research frameworks informing the Carnegie pathways. *NADE Digest, 9*(1), 30–39.

Falk, A., Kosse, F., & Pinger, P. (2020). Re-revisiting the marshmallow test: A direct comparison of studies by Shoda, Mischel, and Peake (1990) and Watts, Duncan, and Quan (2018). *Psychological Science, 31*(1), 100–104.

Fugard, A. J. B., Stewart, M. E., & Stenning, K. (2011). Visual/verbal-analytic reasoning bias as a function of self-reported autistic-like traits: A study of typically developing individuals solving Raven's advanced progressive matrices. *Autism, 15*(3), 327–340. https://doi.org/10.1177/1362361310371798

Jolliffe, T., & Baron-Cohen, S. (1997). Are people with autism and Asperger syndrome faster than normal on the Embedded Figures Test? *Journal of Child Psychology and Psychiatry, 38*(5), 527–534.

Mischel, W. (2014). *The marshmallow test: Understanding self-control and how to master it*. Random House.

Piechowski, M. M. (1997). Emotional giftedness: The measure of intrapersonal intelligence. *Handbook of Gifted Education, 2*, 366–381.

Shoda, Y., Mischel, W., & Peake, P. K. (1990). Predicting adolescent cognitive and self-regulatory competencies from preschool delay of gratification: Identifying diagnostic conditions. *Developmental Psychology, 26*(6), 978.

Torrance, E. P., & Dauw, D. C. (1965). Aspirations and dreams of three groups of creatively gifted high school seniors and a comparable unselected group. *Gifted Child Quarterly, 9*(4), 177–182.

White, H. A., & Shah, P. (2011). Creative style and achievement in adults with attention deficit/hyperactivity disorder. *Personality and Individual Differences, 50*(5), 673–677.

8

Neurodiversity and Collaboration

Jay—Just a Matter of Time

In 1995, I was called to visit a local high school in Halifax, Nova Scotia, where two science teachers had just won a set of webcams for the computer lab. The school's network didn't have the bandwidth for web-enabled cameras, and the teachers were trying to think of different ways to use the cameras in their science instruction. While we were brainstorming ideas, the biology teacher told me this story.

His class was studying genetics, and he found a recent article about the risks and merits of genetic testing. He asked his class to read the paper and prepare their thoughts on the topic. They would each sit in front of a computer and present their argument to the video camera, and later, the class would go through them all. The videos he received, for the most part, were the self-conscious, nervous, giggling teenage productions one might expect.

One video, however, stood out.

Jay was a boy who rarely said a word in class. His teachers had very little success in getting him to demonstrate any progress towards the expected outcomes in class. In other words, he was failing. After turning on the camera and being asked to discuss the article, Jay sat in front of the camera in absolute silence for 42 minutes. Forty-two long minutes of nothing.

In the 43rd minute, however, Jay looked up directly into the camera and started his presentation. Jay gave, by far, the most profound and poignant argument of the class. He explained how his own younger brother had been born with an awful childhood disease and had lived for only a few months. Jay explained the great trauma this event brought to his family. He wondered aloud what it would have been like if his brother had lived. In the remaining 20 minutes, Jay weighed many possible positive and negative impacts that genetic testing would have had on his family if it were available at the time. Using clear examples and accurate information from the assigned article and other studies, Jay explained the complexities of the medical procedures his baby brother had to endure in his short life and the difficulty of the decisions his family had to make.

Everyone was astounded. When given the time, this boy—from whom no one expected anything—provided an eloquent, scientifically based, emotionally charged argument for genetic testing. No one in the school had known the story about Jay's little brother. No one in the school thought of Jay as capable of such profundity. No one had ever given him 42 minutes of silence to gather his thoughts. For some learners, it may take that long, but it can be worth it in the end.

Communication

Clear and precise communication is essential in STEM problem solving. While neurodivergent thinkers may bring the advantage of viewing a problem differently and see new solutions, they sometimes may not be able to communicate that vision in a way others are expecting. Many STEM managers reported finding hurdles in communication as neurodiversity increased in their companies, but they also note that when they overcame those hurdles for a specific group of employees, it generally turns out improving the workplace for everyone.

The American Psychological Association (APA) divides communication into three components: speech, language, and pragmatics. *Speech* comprises everything that goes into producing sounds. *Language* is the meaning of words and how we put them together. *Pragmatic language* includes nonverbal nuances such as understanding gestures, nonliteral communication (such as metaphor, irony, and sarcasm), and detecting the emotional meaning from body language and facial expressions.

The types of communication differences most prevalent in neurodiversity have to do with *social communication*, which is the comprehensive form of communication between people and includes social cues and norms that go along with interpretation of what is communicated. Differences in some nonverbal aspects of communication, such as facial expressions and the tempo of speech, may account for what others perceive as "awkwardness" in people with autism. Sometimes, neurodivergent thinkers find the pacing of conversation hard to follow, not seeing the same clues as others about when to start and stop talking. This can lead to what others see as cutting in or dominating the conversation when they don't mean to. Noted differences in communication can include both pragmatics and prosody. *Pragmatics* is the appropriate use of language in social situations, while *prosody* is the rhythm of speech. In addition to struggles to stay on topic and take turns in a conversation, some learners may be challenged to choose appropriate questions and tone of voice suitable for different circumstances. Differences in prosody result in some learners using monotone, while others may have exaggerated pitch. These changes in tone can be misinterpreted as limited or exaggerated emotion.

Children with ADHD may be more likely to get off topic when speaking, or struggle to find the right words and put thoughts together coherently in conversation. Errors in grammar as they compose sentences also may occur. This may be caused by differences in their ability to plan and parse words rather than any limited understanding. In ADHD, listening comprehension can be impaired when managing distractions. Crowded, noisy rooms are particularly difficult when the person must attend to one conversation amidst many others. Distraction may cause a learner to miss details in conversation and stories. These limitations may be interpreted as oppositional behavior when a request appears intentionally ignored instead of not being heard in the first place. People with ADHD may have to work hard to avoid blurting out answers, interrupting, talking excessively, and speaking too loudly. They also may make seemingly tangential comments in conversation, while they struggle to organize their thoughts. They may understand and respect the common social rules of discourse, but distractibility and impulsiveness may cause lapses in pragmatic skills and thus may undermine social communication.

Difficulties in communication appear to be more pronounced when neurodivergent people are interacting with non-neurodivergent people, or at least with people who communicate differently than themselves. Andrew Begel notes that it is sometimes hard for neurodivergent problem solvers to convince others of their good ideas because they have difficulty getting their point across. This may be because it is hard for some neurodivergent thinkers to imagine how non-neurodivergent thinkers will perceive a situation. The reverse is also true; it is often hard for non-neurodivergent thinkers to predict and understand how a neurodivergent thinker will react or think. This was demonstrated in a study where participants were asked to play a game of "telephone" where a message was passed secretly from one participant to the next, and the final answer is compared to the original communication for accuracy. It was found that communication within a group with only autistic people was just as accurate as in groups with only non-autistics. The message was degraded in transmission, however, when non-autistics played with autistics, showing that the differences in communication style, social–cognitive characteristics, and experiences go both ways. It is, therefore, just as incumbent for non-neurodivergent people to learn how to communicate with neurodivergent people, as it is the other way around.

Speech therapists and other specialists use a variety of methods to support communication with nonverbal learners. One common method is *augmented and alternative communication* (AAC). There are many different types of AAC including technology such as a tablet and/or a computer with a speech-generating device. Non-technological options include including gestures and facial expressions, writing, drawing, pointing to letters to spell words, and/or pointing to pictures or words to communicate ideas.

Another interesting, but much more controversial method, is *facilitated communication*. Facilitated communication is based on the presumption that some nonverbal people want to communicate but may suffer from *apraxia*, which is the inability to do what you want to do at the moment you want to do it. In particular, apraxia of speech affects the brain pathways involved in planning the sequence of actions involved in producing speech. The brain knows what it wants to say but cannot properly plan and sequence the required speech sound movements. In facilitated communication, this apraxia is said to be mitigated by touch and assistance at the keyboard, which helps the learner overcome motor control and/or proprioceptive difficulties that may keep them from typing what they are thinking. *Proprioception* is the ability to sense one's own movement, action, and location in space. In facilitated communication, a therapist typically places a hand on the arm or back of the learner to calm their twitchy movements and ground them in space so that they have an easier time typing what they want to communicate. The APA strongly denounces the facilitated communication method, saying there is no evidence of its effectiveness and even saying it can be dangerous for some learners. Meanwhile, however, supporters of facilitated communication report amazing breakthroughs with nonverbal learners, with adult autistics finally able to express themselves with deep thoughts and emotions after decades of silence.

So, what are you thinking?

1. What is an example from your classroom, workplace, or daily life where communication was complicated by differences in communication?
2. What are ways you can make your classroom or workplace more flexible to allow different modes of communication?

Empathy

Just as clear communication is a key part of collaborative problem solving, so is empathy. To be effective, a problem-solving team must come to a shared understanding of the questions being explored, what needs to be done, and what information is known. By seeing a problem through the perspective of others, problems-solvers are more likely to consider alternative approaches, some of which may lead to better solutions.

The relationship between empathy and neurodiversity is complicated and controversial in the literature. For example, some research suggests that it may be difficult for some autistic people to perceive or detect the cues to understand the perspective of others, but there is other research that argues that autistic people may more empathetic than others, yet unable to express that empathy. Some believe this is a great source of frustration that can lead to outbursts or tantrums. Baron-Cohen has proposed an *empathizing–systemizing theory* (E-S), which classifies people along two independent dimensions—empathizing (E) and systemizing (S)—to establish their Empathy Quotient (EQ) and Systemizing Quotient (SQ). According to this theory, people with autism are associated with below-average empathy and average or above-average systemizing. However, other research found no evidence of impairment in autistic individuals' ability to understand other people's basic intentions or goals. This discrepancy may be explained by not differentiating between cognitive and affective empathy. While *cognitive empathy* is the ability to understand and relate to another person's mental state, *affective empathy* is the ability to respond appropriately to that other person's mental state, which requires more than just awareness and understanding.

Theory of Mind and the Double-empathy Problem

Psychologists have used the term *theory of mind* to describe a person's capacity to understand another person by attributing thoughts, feelings, and perspectives to the other person. Theory of mind is related to empathy, but the distinction among theory of mind, empathy, and perspective-taking remains muddled in the research. The theory of mind has become controversial, especially when interpreted as implying empathy is diminished capacity among neurodivergent people. In reaction, Milton proposed the double-empathy problem, suggesting that the apparent lack of empathy associated with autism is more of a lack in communication and social interpretation, and that this deficit is a two-way problem. As noted in the last section, in a game of telephone where a message was transmitted from one player to the next and the final message was compared to the original, autistic people were just as accurate as the non-autistic group when the people they were playing with were other autistics. The double-empathy problem recognizes that social and communication difficulties between autistic people and non-autistic people can be caused by a reciprocal lack of understanding, not a deficit of autistic people.

So, what are you thinking?

1. In what ways has an awkward communication been interpreted as a lack of empathy in your classroom, workplace, or daily life? Upon reflection, is there another way to interpret the event?
2. How can better communication and empathy be fostered in your classroom, workplace, or daily life?

Carleton—A Missed Opportunity

My first job after college was with the Federal Systems Division of IBM, a contractor for NASA's Johnson Space Center. I was a verification analyst on the Onboard Flight Software for the first 25 missions of the Space Shuttle. Yes, it was the coolest job ever. As a 22-year-old, I was driving around in a new car and a snazzy suit with a pager that could beep any moment, indicating a computer glitch was detected while the astronauts were up on a mission. If the glitch occurred in a section of the code I had been working on for the mission, I was called into a NASA press conference where I had to explain that the computer's error signature looked exactly as it had in the simulation we'd run on the ground, and that we could all be sure the redundancy of the system was working as it should—the astronauts and the mission were safe.

In the early 1980s, professional software developers had the practice of measuring the quality of their code by the number of errors in 100,000 lines. The US software industry standard at the time was 5 errors per 100,000 lines, and the Japanese industry standard was at 3. The Space Shuttle software had less than 1 error per 100,000 lines of code. I was working with the most detail-oriented perfectionists on the planet. (Note: this was before the Challenger disaster in 1986, and a further note, that explosion was not a software error.)

The people I worked alongside during that time were predominantly white and predominantly male. I was used to that. I had been a computational math major with a minor in physics in the 1970s. There weren't a lot of women in my college classes either. Clear Lake City, Texas in 1983, however, also still held the remnants of a culture honed by early test pilots where the cool and silent type were revered, and the quirky personalities were not limited to astronauts. I worked with an extremely talented group of computer programmers and aerospace engineers at IBM. Our job was to create and test the navigation and systems operation software for the space shuttle. Lives depended on our systematic, creative, and persistent problem solving. But the social dynamics were unlike anything I'd ever experienced.

As a junior team member, I had been assigned a project on a small piece of software created for the specifications of an upcoming mission, and I needed guidance from the person who wrote the original code. I was scheduled to meet every Wednesday morning with a man I'll call Carleton. Carleton was at the very top of the technical ladder of IBM. He had single-handedly built much of the verification software architecture for the shuttle program. If you couldn't figure something out, you went to Carleton.

Carleton was also one of the few men in the office who still wore the IBM standard uniform—a blue pinstripe 1950s style suit and white Oxford button-down shirt—albeit his shirttail was often untucked, and black wingtips were perpetually scuffed. The plastic pocket protector lining the left front pocket of Carleton's suit jacket was always full and ready with five different colored pens, each with its own use. My code was constantly marked up with green for logic and blue for variable declarations. Red just meant "start over."

There were no family photos around his Carleton's office, and he rarely came to the after-work social functions that were abundant and popular amongst the IBM staff. As far as I could tell, while Carleton had garnered the utmost respect for his intelligence and programming prowess, he had very few "friends" among the people we worked with.

I was often unrested in those early Wednesday morning meetings with Carleton, having usually spent the previous night tossing and turning in angst. You see, Carleton intimidated the heck out of me. When he reviewed my work, he immediately found errors—subtle things I hadn't noticed or even thought to look for—and he blurted them out in an abrupt manner that left me feeling scolded and ashamed.

I longed to become more efficient, and to become more thorough, if for no other reason than to make my weekly meetings with Carleton more tenable. Each week, I would show him my code, pointing out places where I was getting stuck, and Carleton responded with rapid-fire explanations of another esoteric example and solution, as if I were an idiot not to pick on the intricate connections he saw between the two problems. Not only was I far too novice to understand the depth of Carleton's ideas, but the impatience I perceived in his machine-gun-fire tone stopped me from ever asking any clarifying questions.

The communication was so poor between Carleton and me that often he would just end up finishing the task himself, leaving us both exasperated. To me, Carleton appeared absolutely unwilling to explain his reasoning for choosing his examples or explain to me where or why I was making mistakes. I left deflated, furious, and convinced that he was a belligerent, arrogant man whose intent was to put me in my place.

These days, I often look back at my work with Carleton and wonder if there were better ways we could have worked together. I wonder if there was a different way for me to learn all he had to teach me. If I had recognized his lack of communication as a struggle rather than an attack, could I have found different ways to ask questions? Carleton was possibly one of the most brilliant programmers I have ever known, and I had the precious opportunity to be mentored by him, but I didn't know how to make it happen. And neither did he.

Bibliography for Chapter 8

Crompton, C. J., Ropar, D., Evans-Williams, C. V., Flynn, E. G., and Fletcher-Watson, S. (2020). Autistic peer-to-peer information transfer is highly effective. *Autism*, *24*, 1704–1712. https://doi.org/10.1177/1362361320919286

Frith, U. (1989). A new look at language and communication in autism. *International Journal of Language & Communication Disorders*, *24*(2), 123–150.

Green, B. C., Johnson, K. A., & Bretherton, L. (2014). Pragmatic language difficulties in children with hyperactivity and attention problems: An integrated review. *International Journal of Language & Communication Disorders*, *49*(1), 15–29.

Milton, D. E. (2012). On the ontological status of autism: The "double empathy problem". *Disability & Society*, *27*(6), 883–887.

Rogers, K., Dziobek, I., Hassenstab, J., Wolf, O. T., & Convit, A. (2007). Who cares? Revisiting empathy in Asperger syndrome. *Journal of Autism and Developmental Disorders*, *37*(4), 709–715. https://doi.org/10.1007/s10803-006-0197-8

//

9

Inclusive STEM Teaching

A Classroom Observation—This Hour Has 22 Minutes

There are many aspects of today's schooling that do not lend themselves to learner-centered learning. The amount of time spent in the classroom on things other than learning has increased tremendously over the years. It is hard to pin this on one change. It is exacerbated by compounding changes in administrative recordkeeping for attendance and student support, testing, security issues, parental involvement, and overall student needs. The result, in my observation, is a decreased opportunity for rich, uninterrupted learner-centered learning experiences. To illustrate this point, here is a set of my observation notes from a grade 7 class I visited for observations in 2015.

10:57 am.

- End of Third Period bell rings.
- Thirteen classes of children cram together in the corridors in a 5-minute fury to get to their next classes.

11:02 am.

- Start of Fourth Period bell rings.
- Twenty-two children are in the room.

- Eight are sitting in a chair with their jackets off, notebooks out, and quietly (if not enthusiastically) awaiting next instructions.
- Four are at or sitting individually in a chair with their jackets and backpacks still on or not out away.
- Nine are clumped in small groups, standing and sitting, chatting as if they are not awaiting anything to happen.
- One is in the back of the room, digging through a carton of packing material that had recently been used for shipping.
- Teacher is taking attendance and writing slips for the 5 students missing from class.
- Teacher asks a student (who was waiting patiently for class to start, ready to engage) to take the slips down to the office. (While the last slip is being prepared, one of the missing students arrives. The teacher must sort through the slips and retrieve it so that the correct slips go to the office.)

11:07 am

- Student leaves to take slips to office.
- Another tardy student arrives. Teacher completes slip to take to office later to correct attendance.

11:11 am.

- The teacher projects an introductory slide and asks the class to settle down.
- Another tardy student arrives accompanied by an educational aide. Another slip is filled. A chair is moved from across the room to give the aide a seat.

11:13 am.

- The teacher tries again to focus the class on the introductory slide. Teacher introduces the topic of Clear Commands (how to talk with a computer) and tells the class that they are going to work in pairs.

- Students are looking around for partners, a small din of conversation builds.
- Teacher tries again to focus students on the activity. Teacher explains that students will create a set of commands that their partner will use to solve a maze. They will each create and follow directions so they can each test each other's work.
- Teacher asks if there are questions. Six hands go up.
- One student asks if they have to work with the partner who chose them.
- One student asks if they can go use the computers in the computer lab.
- Two students ask if they can go to the bathroom.
- One student asks whether the teacher received a previous assignment left on the teacher's desk.
- One student asks to go to the office to clear up paperwork for the after-school basketball trip.
- The student who brought the tardy slips down to the office re-enters the classroom. They ask the girl next to them for an explanation of what they missed.

11:18 am.

- The teacher returns to the introductory slide and is finishing up instructions when the overhead PA systems chimes. An announcement about the after-school basketball trip comes on. The instructions on where to meet the bus and what to remember to bring is followed by a list of students who are requested to come down to the office to complete unfinished paperwork for the trip.
- Three additional students get up to leave to go to the office.

11:20 am.

- The teacher instructs the class to assemble the supplies so that they can start the maze task

- Students mull around to gather a pencil and a maze worksheet (students elect one of three choices from the front of class), and a partner to work with.

11:24 am.

- Student returns from the bathroom. Teacher settles student in new group and asks group to give instructions.
- Six pairs of students are working together on the maze activity.
- Two other pairs have chosen to work together, but one pair is engaged in a game on a phone under the desk and the other is braiding her partner's hair.
- Three students are still mulling about gathering materials or disengaged in their chairs, not attempting to find a partner or start work.
- Teacher is called to the front of the class to assist a student who has returned from the office to retrieve their lost password to the classroom management system.

11:31 am.

- Three students return from office and one from bathroom. Teacher introduces them to activity.
- The teacher circulates through the class, getting an overview of what each group is doing.
- Six pairs have finished the mazes and are talking about them.
- Two pairs are writing down the commands for each other to follow (as instructed).
- One pair and four other students are working independently on mazes.
- Four students are completely off-task.
- Teacher is sidetracked by a conversation about a personal matter with a student.

11:38 am.

- Teacher asks who has completed the maze commands and can demonstrate for the class.
- Three pairs of students offer to demonstrate their completed activity.

11:47 am

- Lively and productive discussion engages students who completed the activity and those who didn't.
- Student who did not complete maze offers suggestion to improve the commands for another pair's activity. Students wants to go back and try again.
- Discussion continues among students as an aide starts to gather the belongings of the three students who must leave early for a supported lunch program.
- The teacher rises to capture key concepts from the discussion on the board and to solicit next steps, but students are already putting away their binders. No notes were taken.

11:52 am.

- The end-of-class bell rings while the Teacher yells over it saying they will continue the discussion on the maze during the next class, which is a week later.

I started these observation notes with the intent of examining engagement with the maze activity to understand how to make the activity better. What I came away with was the understanding that what we designed as a 45-minute activity had to be crammed into only about 20 minutes of actual teaching and learning time. With the number of interruptions, disruptions, and differentiated needs in today's typical classroom, it is nearly impossible to accomplish an entire learning cycle—introduction, learning activity, reflection, and discussion—about anything. It reminds me of the CBC television show called *This Hour Has 22 Minutes*, referring to all the time lost to commercials in a network TV hour.

We expect learners to focus their attention, absorb new ideas, and think critically within these fragmented 20 minutes, and then start it all over again during the next period. By the time most of these students revisit this activity a week later, it will be a distant memory with little carryover. It is an absurd way to try to teach and has very little grounding in research about how people learn.

Neurodiversity in Schools

In the US, the American Disabilities Act (ADA) enacted in 1990 gave civil rights protections to individuals with disabilities, guaranteeing equal opportunity for public accommodations, employment, transportation, state and local government services, and telecommunication. The Individuals with Disabilities Education Act extended the rights of the ADA to give each child the right to accommodations they needed to pursue the mainstream curriculum in education. In 2001, the No Child Left Behind (NCLB) modification to the Early and Secondary Education Act (ESEA) places a stronger emphasis on accountability including academic content standards, achievement standards, and statewide assessments. NCLB was also intended to increase flexibility and local control, extend options for parents, and emphasize teacher qualifications and methods.

In the learning science community, many of us saw NCLB as highly limiting and contrary to the student-centered learning approaches we were trying to promote. NCLB's focus on statewide testing seemed to result in teachers "teaching to the test," and our learner-centered reform efforts felt sidelined. NCLB felt like a major roadblock to the learning science movement. Many administrators and parents, however, saw NCLB as a chance to bring in standards and accountability, providing a coherent avenue for school improvement.

In my conversation with Nadine Bonda, a former math teacher, principal, and superintendent in the Massachusetts education system and also the chair of TERC's board, she stressed the beneficial impacts of NCLB for the improvement of teaching and learning. Bonda explained: "Before NCLB there were no standards. There was no way to hold teachers accountable that made any sense. The development of standards and NCLB were inextricably linked. With NCLB, school districts had the goal to ensure each of the children met a specific set of educational outcomes and therefore the districts had to start getting serious and systematic about curriculum development."

Bonda notes that the development of standards played out differently in different districts and in different states. She said that in Massachusetts, there was ample input from teachers and researchers into the standards with the Department of Education coordinating the effort, and she felt they came up with pretty good standards that allowed teachers room for creative and differentiated teaching. As a result of NCLB, Bonda feels many districts now have the tools to be able to talk about what good teaching looked like. She says that through NCLB, the question on the district level became: "How do you improve instruction in school so that students are effectively able to meet the standards?" With this shift towards bringing all learners up to a certain level of standards, Bonda says differentiated instruction has come more to the fore, expanding the idea that different children in the classroom require different types of supports to achieve the prescribed standards. Bonda said other states may not have been as fortunate.

The impact of NCLB on neurodivergent learners is also a mixed bag. Before NCLB, there was some standardized testing in schools, usually in reading comprehension, but it was not closely tied to the curriculum. Some neurodiversity advocates had observed that neurodivergent students were being asked to stay home on test day so they wouldn't be counted as part of school assessments. NCLB required the inclusion of all students and was intended to get a better read on how each child in the school was progressing. Some parents believed that their neurodivergent children were finally being recognized as part of the larger system and were provided with supports. NCLB also stated that every child would be served within the least restrictive environment possible to meet those outcomes, and that each child would have the right to the services needed to guide them to those outcomes within that environment. This started to shift attention to inclusive education.

In 2015, President Obama signed the Every Student Succeeds Act (ESSA) into law, saying:

> The goals of No Child Left Behind, the predecessor of this law, were the right ones: High standards. Accountability. Closing the achievement gap, but in practice, it often fell short. It didn't always consider the specific needs of each community. It led to too much testing during classroom time. It often forced schools and school districts into cookie-cutter reforms that didn't always produce the kinds of results that we wanted to see.

The ESSA provides more flexibility for states to devise their own standards and testing plans. ESSA emphasizes the need for measurement other than just test scores (e.g., including graduation and dropout measures) and the inclusion of mid-year tests to inform teaching.

The flexibility and equity emphasized by NCLB and ESSA has brought a rise in the number of Individualized Education Programs (IEPs). Similar tools are used in Canada (e.g., in Nova Scotia they are called Individual Progress Plan and in the UK they are Individual Education Plan). An IEP is a plan arranged among the school, parents or guardians, and learning specialists to provide accommodations and supports for the student in school. Today in the US, about 15–20% of students in public school have an IEP or equivalent. An entire industry has grown up around advocating for rights under IEPs in parts of the US. Parents can hire a special education advocate to help them work with education systems to get the supports they consider their child is due. There are also special education law firms that specialize in litigation to enforce the provision of education their child is entitled to by law.

Parents who are savvy and well connected in the community— better educated, English-speaking, and more aware of what the school has to offer—are often in a better position to advocate for their children. On the other hand, parents who do not have financial resources for these services, or don't speak the language, or who went to school in different education systems may not know what schools can offer and are not likely to benefit from these services. Those who are struggling to raise children while holding down two or three jobs may not have the time or resources to advocate for their children. So while the educational system may have been designed for equity, the ideal has far from become realized.

In Canada, there is a greater trend towards inclusive classrooms where neurodivergent students spend more time in general education classes with educational assistants. While there are many benefits to this approach, it has to be well resourced with well-prepared teachers and assistants to ensure a productive experience for all involved. Learner-centered learning in inclusive classrooms call for differentiated teaching and learning materials and flexible learning environments that are designed to be adaptable for a broad range of learners.

So, what are you thinking?

1. What policies and practices affect the inclusion of neurodiversity in your school, workplace, or daily life?
2. In what ways have changing policies allowed more or less support for neurodivergent learners in your classroom, workplace, or daily life?

Differentiation and Universal Design

Inclusion is the idea of educating learners together in one classroom (or one learning environment) without separating them by needs. Inclusion may mitigate some of the disruptions when learners are "pulled out" to meet with specialists. But inclusion is not enough—meaning that putting all learners together in one classroom does not, in and of itself, guarantee inclusive learning. The design of the environment and learning activities must be specifically designed for inclusion. For effective inclusion, teaching and learning experiences need to be *differentiated* for each learner, adapting to the unique physical, cognitive, emotional, and social considerations for each learner. Most general education and inclusive classrooms in elementary school have 20–30 learners with one (or maybe two) lead teachers. There may be a paraprofessional or two to support learners with IEPs, of whom there are likely at least about four or five in the class. Middle and high school usually have more students per class and fewer supports. Differentiating for this number of students is a dauntingly challenging task.

CAST and UDL Guidelines

This is where design and technology play a role. In 1984, the same year Apple released the Macintosh computer, a group of advocates for people with disabilities met outside Boston, MA to discuss how technology could be used to provide better educational experiences for people with disabilities. This group evolved into the Center for Applied Special Technology, which later became known as CAST. CAST and Apple worked together in those early years to make use of Apple's new graphical interface and built-in text-to-speech features to provide access to more learners. Since then, CAST has grown into a world-renowned educational group that has set a framework for Universal Design for Learning (UDL), the de facto standard for design principles for accessibility in education. UDL focuses on differentiation teaching and learning activities in three areas: engagement, representations, and action and expressions. Fostering engagement starts by assessing learners' knowledge, interests, and skills so the learning experience can be catered to those motivating elements. Often this requires establishing a personal relationship between teacher and student, so that areas of relevance and sensitivity can be revealed and learning experiences can be differentiated accordingly. This, of course, is very difficult to scale to a large number of students.

The UDL guidelines emphasize that while some learners may input information faster if it is visual or audio, no one single means of representation will be optimal for all learners. Multiple representations offer more opportunities for engagement, and repeated application of a context may also allow students to reinforce the knowledge and make connections across settings. CAST additionally suggests that when students are allowed multiple means for expressing and acting upon their ideas it will lead to more inclusion of diversity in high-quality learning experiences. Some learners may express themselves most fluidly in writing, others verbally, and others pictorially. Some students may also need extra supports to communicate their understandings in any form, and this is where technology aids such as speech-to-text translators can be particularly effective. All of these can be seen as valid ways to learn and communicate student understanding.

Flexibility and Adaptability
The emphasis on choice and multiple representations in UDL is aligned with many aspects of learner-centered learning and constructivism, yet it also presents a challenge in work with neurodivergent learners. Along with choice and multiple avenues of entry comes complexity in how the experience is delivered, and the potential to increase the extraneous cognitive load for the learner. For some learners, just sifting through the various options and making a decision is taxing and takes their attention away from the task at hand. Therefore, the balance of flexibility with the simple, clean interfaces is a holy grail in design for neurodiversity.

For this reason, there are also many efforts using machine learning and artificial intelligence in adaptive learning environments to try to replicate (or possibly surpass) this one-on-one interaction so that differentiation can happen at a large scale. This research is still young, and many agree that even when learning analytics can be used to customize cognitive tasks (such as a personalized math tutorial), the social relationship that motivates learning for many students may not come from a machine. I, too, believe there will always be an important role for teachers—especially excellent teachers—but I also think technology can make this job a little easier. Game-based learning is an area particularly ripe for this type of adaptive learning and differentiation and will be discussed more in Chapter 11.

Differentiated Assessments

The UDL principles of design are also important when designing learning assessments. Because differences in learning can vary over time, teachers require frequent formative assessments to ensure their teaching methods are able to differentiate for each child and keep up with changes in what they need. Unfortunately, many learning assessments often include irrelevant barriers, such as text or other symbolic notation interpretation that may mask conceptual understanding for some learners.

Several years ago, I attended a research conference on dyslexia and dyscalculia in STEM. A number of innovative researchers got up to present research on their interventions to help kids with dyslexia learn mathematics. The interventions were great, but when it came time to study the efficacy of the intervention, their studies used complicated word problems to measure outcomes. That meant that learners with dyslexia had to parse out a paragraph of text and recreate someone else's imaginary scenario in their head before they could even start tackling the math problem. I wonder how many talented math learners were missed in that research.

Inclusive assessments should be specifically designed to highlight the strengths of learners, and be able to be flexible enough for differentiated teaching and learning to nurture those strengths while also supporting learners' unique needs. Learning assessments that adhere to UDL principles require both the delivery of the question or task and the receipt of the response in the context and modality that best enables the learner to demonstrate what they know. This type of asset-based, differentiated assessment swims against the tide of standardized measurements but may be essential to supporting neurodiversity in the classroom. In the next few chapters, I will show a few examples of what this type of inclusive teaching, learning, and assessment might look like. These ideas are nowhere close to magic bullets; they are fledgling solutions. I hope they stimulate more design, development, and research in this area. There is much more to learn.

So, what are you thinking?

1. What types of differentiation do you already do with your students, employees, or other learners in your life?
2. What areas could you be differentiating more? How?

Jen—Getting to Completion

Jen was a well-behaved student in Ms Bradbury's science class. She was popular with many of the students in her grade, she was always attentive in class, and she was well organized. She didn't get mixed up in the disruptions that sometimes derailed Ms Bradbury's lessons. Jen rarely raised her hand, but when called upon, her answers were always well conceived and showed a solid understanding of the material being discussed. Basically, Jen was the kind of student no one worried about.

Jen's grades, however, began suffering in grade 8. She had always been a bit slower getting her ideas onto paper, but over time she stopped turning anything in at all. When Ms Bradbury questioned Jen about her work, she said it wasn't yet good enough to hand in. She was afraid of getting something wrong. She wanted to hold on to the work until she could make sure it was perfect and worth top marks. This tendency became a concern in her other classes as well. Meanwhile, Ms Bradbury and other teachers noted Jen's chewed-off fingernails, which were at the point of bleeding, and her overall pallor and lethargy. By winter, Jen's anxiety and lack of assigned work completion became unmanageable. She was slipping fast.

Ms Bradbury set up daily meetings with Jen. She set her up with an organizer—a daily agenda in the colors Jen liked—and coached her through some scheduling strategies. Together, they broke up Jen's to-do list into a set of smaller tasks and slotted them into the times when Jen thought she could get them done. There were further meetings scheduled with Jen and her parents. They set up some strategies to work on at home. Jen wanted to do better and tried. She was engaged in class discussion again, but in the end, she still would not turn in her work. Ms Bradbury continued to check in with Jen regularly for the next few months, but none of these tactics seemed to give Jen what she needed to prevail. And by then, her confidence was also shot.

Before the end of the year, Ms Bradbury and Jen's math and English teachers worked together to make some adaptations to Jen's schoolwork by explicitly breaking down her tasks and sticking to hard deadlines for smaller pieces of the work. More importantly, they took the focus off the final product. For example, during class, teachers asked Jen explicit questions about the assignment she was working on. They asked her to explain what she was thinking, her interpretations of the material, and her plans for her next steps. Jen excelled at that kind of discussion. It was clear she had all the content in her head. It just got stuck in there and she was having trouble getting it out.

Jen's teachers assessed her progress in class and asked to see a draft of what she had each few days. They assured her that they knew it wasn't final, they just wanted to see that she was working on it. There would be no evaluation of the product itself. She was okay with this, and they used her drafts as conversation pieces. Her teachers gave her grades on her content from the discussion. With the determination of a grade out of the way, the act of creating the end product wasn't so intimidating for Jen. Her teachers had already seen that she knew the material, and she knew she could not improve her grade by ruminating on it anymore, so she was able to finish the assignment and turn it in. This became a successful assessment strategy for Jen, and may work for others whose anxiety stands in the way of their completion of a task.

Bibliography for Chapter 9

Americans With Disabilities Act of 1990, 42 U.S.C. § 12101 *et seq.* (1990).

CAST. (2011). *Universal design for learning guidelines version 2.0.* Author.

CAST. (2018). *Universal design for learning guidelines version 2.2.* http://udlguidelines.cast.org

Hehir, T., & Katzman, L. I. (2012). *Effective inclusive schools: Designing successful schoolwide programs.* Jossey-Bass.

No Child Left Behind Act of 2001, 20 U.S.C. § 6319 (2008).

Skinner, R. R. (2019). The elementary and secondary education act (ESEA), as amended by the Every Student Succeeds Act (ESSA): A primer. *CRS Report, 45977.*

Tomlinson, C. A., & Strickland, C. A. (2005). *Differentiation in practice: A resource guide for differentiating curriculum, grades 9–12.* ASCD.

Tomlinson, C. A., & Tomlinson, C. A. (2017). *How to differentiate instruction in academically diverse classrooms* (3rd ed.). ASCD.

10

Strategies for Inclusion— Project-based Learning

What Is Project-based Learning?

The term project-based learning (PBL) has been kicking around in the education community for decades. Some educators also use PBL to mean problem-based learning. Not dissimilar, both project- and problem-based learning situate learning in a context, rather than focusing on isolated skill building. In PBL, projects may range from a set of confined tasks introduced by the teacher to an open project for which the course of action, plan, and design is not determined by the teacher. Projects may involve others in the community.

PBL is thought to enhance students' motivation. Students may be personally interested in solving a specific problem, particularly if relates to their daily life or their community. They may also use PBL as an entrance point into deeper study of related phenomena, situating their learning within a real-life application. These types of learning experiences put the learner in the driver's seat, providing agency and autonomy, and often embrace the strengths of neurodiversity. When people learn to create, their learning matters to them. It is worth their efforts to persist, even when that means overcoming other struggles. When learners are allowed to think and express themselves in ways that make sense to them, they can build and demonstrate knowledge, sometimes far surpassing what others have expected from them.

John King, the former U.S. Secretary of education, describes the affordances to reach marginalized learners.

> Sometimes folks look at low-income students, students of color, English learners, students with disabilities and say to themselves, "Because they are behind in some way academically, I'm going to make a very drill-focused, sort of minimalist curriculum." And that is exactly backward. The way that we're going to accelerate the students who most need support is through rigorous, engaging learning experiences like project-based learning when it's done well.

The George Lucas Foundation (GLF), which has studied PBL extensively, states that "One of the principles of rigorous PBL that drives equitable learning experiences is the emphasis on authentic learning contexts, meaning learning experiences should be authentic to students' identities, interests, and communities." GLF also explains: "Teachers must work to embrace asset-based teaching and interrupt deficit thinking based on negative stereotypes and assumptions that limit what is possible for some students."

In my conversation with Dr. Joseph Krajcik, who led the design of a large PBL program known as Multiple Literacies in Project-Based Learning at Michigan State University, he agreed that PBL requires designers and educators to engage in asset-based thinking and planning, and this may be a shift from the deficit view often held in education. He says that by giving students a chance to dwell in phenomena they are interested in, and to pursue questions of their own interest, educators allow learners to learn on their own terms. Mitch Resnick, a professor of learning research at MIT, and his colleagues have empowered PBL teaching and learning with digital tools such as Scratch, an introductory programming environment that enables young learners to express themselves through digital creativity. Resnick also advocates for the Maker Movement, which embraces access to tools, technology, and community to enable all people to become creators.

When I started working at Hillside with Ms Bradbury, neither of us was aware there was already a PBL initiative brewing in the back rooms of the Nova Scotia Department of Education. While they were building a framework for PBL teaching and learning for Nova Scotia schools, Ms Bradbury and I also happened to be trying it out in the classroom. We just started doing it because it made total sense as a way to engage her students, and it worked.

In our third year of working together, Ms Bradbury was called down the Nova Scotia Department of Education for six weeks to work with a team of other classroom teachers and provincial staff to design a PBL curriculum for the rest of the schools in the province. Hillside was selected as a pilot school for the new curriculum. The rest of this chapter is the story of that journey.

So, what are you thinking?

1. What types of projects do you like working on?
2. How do you prefer to structure and learn what you need to know to complete those projects?

Project-based Learning at Hillside Junior High

When I first started visiting Ms Bradbury, she taught science using a provincially assigned textbook supplemented with videos, demonstrations, and as much hands-on activity as she could to help her students meet the provincial science outcome standards. She spent a lot of time in front of the class talking to the class and facilitating conversation. Her high-energy enthusiasm for nearly every topic was contagious for about half of her class. The other half of her students were basically checked out.

Ms Bradbury also taught TechEd once a week to the grade 7 students. The provincial outcomes for TechEd were not part of the graduation requirements, so TechEd grades were not high on the radar for students, parents, or the school system. It was the perfect place to experiment. In addition, an enthusiastic new teacher had just arrived at the school, Mr. Kurt Jerrett. Mr. Jerrett taught grade 9 Tech Ed along with social studies and PE. With a heavy new course load, he was thrilled to be supported in TechEd, so that he could focus on preparing for his other classes. For all these reasons, we decided it was a great year to initiate PBL throughout TechEd in Hillside Junior High. The principal gave us enthusiastic support. The grade 8 students also participated in PBL and were part of my research, but since there was no consistent teacher in the role, I do not discuss the grade 8 teachers who rotated through that position.

Gathering Resources and Equipment

To initiate PBL, we curated a bunch of online and offline resources and bought a few supplies so that we had something to start with. Neither of the teachers nor I had much experience, if any, with the tools. We told the kids they could do just about anything project they wanted and that they could work alone, in a pair, or in a small group. We introduced students (and ourselves) to introductory coding programs such as Scratch and Hour of Code, jewelry-making kits and electronic textiles, and Arduino (simple circuit and computing) kits they could use for their projects. We also showed them sites such as Instructables.com with video instructions for DIY projects on a huge range of topics and exciting video clips with projects that might interest them. We told them we could order parts if needed for construction (within reason).

We also showed the students several easy-to-use online programs they could use. These included home-building websites where they could design houses and rooms for different specifications, a music production software program where they could do mash-ups of songs, and a general web development program. We invited a game designer to talk about how to design a video game, and an engineer to show how to design clean water systems. We discussed examples of social enterprises developed within their own community. We did a lot, but we were winging it.

FIGURE 10.1 Students making a glow-in-the-dark phone case in their Tech Ed PBL class at Hillside
Source: Photograph by the author

Setting Milestones

In our first year, we also had very little structure in the student PBL materials. We gave students a set of four project milestones on a worksheet. They were directed to keep these worksheets in a binder with their project notes. The milestones were:

Milestone 1: Project Definition: State the goal and purpose of the project.
Milestone 2: Project Preparation: Identify the supplies and help you will need to complete the project.
Milestone 3: Project Plan for Implementation: List the steps you plan to take to complete the project.
Milestone 4: Project Evaluation: List the criteria you will use to decide if your project is a success.

We started the first few classes showing example projects and tools. After that, most of the class time consisted of students finding their milestone binders, breaking up into work groups and getting organized, and then proceeding on their project plans. The amount of time spent on each of these tasks was highly variable across students. Some students never got organized, while others were delving into their projects productively from the time they arrived in class to the time the bell rang for dismissal (and beyond).

Finding Purpose
When we started, most days I had no doubt that PBL was the right thing to do for these kids. I just didn't think we were doing it well. There was a point where the teachers looked up at their class, who at the moment were all working independently, and said to me joyfully, "Oh my goodness, I'm not teaching!" What they meant was, "I am not standing at the board talking, and my kids are still learning." Many of the kids said that TechEd class was their favorite and that they enjoyed PBL. It had served as a good hook. Watching this transformation take place was incentivizing enough to keep trying to get it right.

I was disappointed in that first year that student projects were predominantly dream houses built in the online design studio and video mash-ups. There was little variation and little innovation past the examples we had shown them. Upon reflection, the teachers and I realized that we needed to provide more scaffolding not only to the doing of the projects but also to the selection and planning of the projects. We had talked with students about innovation, solving new problems, and choosing problems with a purpose, but that didn't mean they were ready to be innovative themselves. It didn't come naturally to many of the students.

We had been emphasizing the word purpose. We wanted the students to choose a project that was purposeful to them so they would have agency, and we wanted them to articulate that purpose, so that they could make purposeful design decisions along the way. But we realized afterwards, that using the word over and over didn't necessarily help the students find their own purpose. This was especially true for students who had never been given the opportunity or hope to reflect on their own agency and desires in learning.

I remember sitting one day with a pair of girls who had seemingly no initiative, no self-confidence, and no desire to do anything. When searching for a possible project topic, I asked one of them, "What makes you happy?" One girl shook her head with no answer. So I rephrased the question. "When was the last time you were happy?" She sadly said, "When I was asleep." I wanted to give up.

But while grappling with how to engage kids who were not accustomed to being engaged, I also started to look around and see what was happening once we got them going. Some of the kids who had never excelled academically were becoming inventors and designers and project leaders. They started coming to class eager and interested to learn. Together, the teachers and I doubled down on PBL and started building the teaching and learning materials that Hillside students needed to build projects that were meaningful to them.

So, what are you thinking?

1. How might you adapt some of the PBL strategies and tools, such as milestone worksheets, to your classroom, workplace, or daily life?

Eli—Giving New Life to an Old Computer

The spring of grade 9 is a hard time to teach at Hillside. Most of the kids are just riding out the last semester of junior high. They have grown too big for the school and already have their eyes set on high school. That is when we first tried project-based learning. When we started PBL in the grade 9 Tech Ed class, which was taught once a week by a first-year teacher, we had signed out the computer lab for each class.

A boy named Eli sat slumped in his chair near the back of the lab, hoodie up over his head, waiting for the hour to be over. He rarely interacted with other students. Eli didn't have an IPP, but he wasn't doing particularly well at school. He was just getting by. While the other kids joked and flirted and tussled with one another, Eli withdrew deeper into his hoodie. If allowed to sleep through class, it seemed he would have been out cold in no time.

My initial conversations with Eli were met with monosyllabic responses and blank expressions as I asked him about project ideas and favorite hobbies. He didn't seem to want to engage on any level. I showed him instructables.com, a site with a bunch of DIY projects meant to spark his imagination, but two class periods went by without a glimmer of interest. On the third week of class, the school janitor, Mr. Bruce, happened to be clearing out a few of the 15-year-old computers that were defunct and collecting dust in the corner of the lab. Eli poked his head out from under his hoodie. "What're you doing with those?" he asked Mr. Bruce. I was startled at the sound of Eli's gravelly voice. "There's a warehouse down by the school board buildings full of old dead computers," Mr. Bruce replied. "I'll probably just bring 'em down there."

Here, I should pause and say something about the Hillside school janitor, Mr. Bruce. He had been at the school for decades. A slight, jolly, and non-intimidating man, Mr. Bruce was beloved. His office on the ground floor of the school, with every speck of wall adorned in memorabilia from his favorite hockey team, was a sanctuary for many Hillside students. It was a place they could always go to hear a simple hockey story or joke to break up the stress of their day. Mr. Bruce was a friend to everyone in the school.

Eli sat up and asked Mr. Bruce, almost jokingly, if he could take one of the computers apart, likely anticipating the answer to be "no." Mr. Bruce looked at me quickly and then responded with a wink, "Yeah, no one down there is gonna miss an old computer anyways." And with that, a project was born.

Eli spent every minute of Tech Ed class, along with ample time during lunch and after school, taking apart that old computer and documenting each component. He cleaned and tightened everything he touched. He catalogued all the computer's components using a system he had devised to track each piece, sorting them by envelopes so he wouldn't lose anything along the way, and looking them up on the Internet to see what they were. At the end of each 55-minute period, he carefully gathered up all the loose pieces and stored them in an organized fashion so that he could come back to the next week and start where he left off. He spent his evenings watching YouTube videos about how computers are built. He created charts to remind himself how everything should be put together, and he came to each class ready to work.

And then he rebuilt the computer. Along the way, he tested each of the components in a separate, functional computer to see if it would work. It reminded me of trying to figure out which Christmas light in a long strand was the one that was burnt out, causing the rest of the string to go dark. In the end Eli, with the help of the school board IT guy who happened to visit along the way, found the video card that was fried and had stopped the computer from working. They replaced the card with another one they salvaged from a different computer, and they fired up the machine. The look on Eli's face when the old Windows 95 logo appeared on the screen was jubilant, something I hadn't known was possible when I'd first met him. He had done it and he was proud.

Supporting Executive Function in Project-based Learning

Having a year or two of PBL under our belts, the Hillside teachers and I thought seriously about how to revamp and refine our strategies. In the first year, we chose a lot of free software and I had personally donated most of the basic supplies. These small contributions soon snowballed into acquisitions of thousands of dollars' worth of equipment for the school. Ms Bradbury and Mr. Jerrett and their student put together a video about their project work and ended up winning a national prize that came with a $10,000 maker space kit. The school also qualified for a district-level pilot program that included robotics kits, software, and a bank of new computers. At that point, the principal let us take over some space in the art room that was adjoined to Ms Bradbury's room to house all this new equipment, and soon we set up a small maker space for PBL in the school. The

FIGURE 10.2 Student using an online art program in Tech Ed class
Source: Photograph by the author

FIGURE 10.3 Hillside Tech Ed classroom with makerspace equipment
Source: Photograph by the author

accolades and momentum after receiving the award, and the students' enthusiastic reactions to PBL, gave the Hillside principal and the teaching staff a boost in confidence about this new endeavor. Each of the TechEd teachers was branching out in new ways to support their students.

Tapping Into Students' Interests

Ms Bradbury realized that building learner agency and tapping into learners' own interests was the area where she needed to work most with the grade 7 students. They were having difficulties choosing projects, and she didn't want to give up on making things personally relevant to them. She developed a "Getting To Know You" questionnaire for her class. While some students offered this information easily, others needed time for serious reflection to figure out their own responses. The students' responses fostered class discussions to outline projects that would be purposeful and meaningful to their own lives.

School Design Project

Mr. Jerrett chose a different way to structure PBL with the grade 9 students. He didn't want the experience to feel like a repeat of their previous year's TechEd experience, and he also felt the small open-ended projects were not connecting with some of his students. He chose instead to engage the entire class in one project—the design of a new school. Public discussions had started about a new junior high school in the neighborhood. A potential site near Hillside was being explored, so we asked the class to come up with a design for a school they would like to see at that location.

Mr. Jerrett brought the class down to the site on a sunny day to walk the grounds, take measurements, and think about their designs. The students conducted their version of a needs assessment of the community by surveying their fellow schoolmates on what was important to them in a new school. They used the Internet to study how those needs had been addressed in other examples. They examined school buildings around the world while they each designed a type of room (e.g., a gym or a music room or the cafeteria). Each group built a scale model of their component, either physically or digitally (or both). Mr. Jerrett coordinated with the math teacher to use the same approach to scale-modelling in Tech Ed as they were learning in math. Two students were chosen to be "project managers" in each class, determining the overall scale for the school model and making sure everyone's components would fit together in the full-school model.

128 ◆ Strategies for Inclusion—Project-based Learning

FIGURE 10.4 School Design Projects by Hillside Grade 9 Students
Source: Photograph by the author

Each (of two) grade 9 classes built a school model and presented their designs to distinguished visitors, including an architect we brought in to support the project as well as city councillors and school board members. The school design project was successful in engaging a large portion of the class and serving as a wonderful integration of math and social studies that students were learning in other classes within their Tech Ed curriculum. Students used skills from their current math classes to create scale models. They created surveys and interview tools to do the needs assessment, discussed the ethics and potential biases of their decision-making processes, and learned Microsoft PowerPoint to create their presentations. They felt their own potential agency and civic responsibilities, as well as gaining public speaking practice.

Supporting Planning and Organization

By the end of the second year, Ms Bradbury and Mr. Jerrett both said they felt they were doing a better job connecting with students. Because the teachers were no longer at the front of the class "teaching," they spent most of the class time circulating among the groups of learners. They were there to answer questions, provide suggestions, and encourage them to focus and persist when needed. While they may not have talked to every student every day, over the period of several weeks, they had substantive conversations with each student more than once. They were also ready to respond when a student exclaimed, "Mr. Jerrett, come look, I did it!" or when they asked, "Ms Bradbury, can you come help me with this?"

As PBL got going in their classes, the teachers had another epiphany. They had already adjusted to the fact that they weren't at the front of the class teaching anymore, but now they realized that they rarely had to think about which kids were on an IPP, and fewer kids were leaving class for special attention. Students who needed to move were free to stand up and walk around if they needed. Students who were non-English speakers used Google Translate to get what they needed. Students who wanted to work alone were able to, and student who wanted to work with their friends were allowed as long as it was a productive relationship. Students who wanted to build could build. Students who wanted to design on a computer could do that. They were each working to their own interests and strengths, and as a result, there were less classroom management or discipline issues.

Still, many students had trouble with the planning and organization of a project. They needed more structure and explicit tools to monitor their own progress. The teachers came up with a set of expanded sheets in the second year, adapting the content to each grade level, but providing a coherent structure through all three grades. The milestone worksheets also listed the TechEd outcomes that could be graded at any time during the project. Finally, we adapted the outcomes and milestones into graphic organizers to help students be metacognitive about their projects and their progress.

Supporting Transitions

We also realized that many of the executive function difficulties kids were facing in their project work happened at the beginning and end of class. Students had trouble getting started with their project work, often getting distracted while they were still getting organized so they couldn't get into the flow of their project work and become absorbed. They didn't take the time to wrap their work up neatly at the end. Transitions were hard. Many times, at the start of a class, when we tried to get students going, we would ask them what they needed to accomplish and they referred back to the overarching goal of their project. They were not breaking the project down into individual tasks that could be focused on each day. Others were focusing so much on the details of an individual task, they were losing sight of how it fit into the bigger picture. To address these issues, we supplemented the milestone worksheets to include daily task sheets (Figure 10.5). Yes, this meant more papers for the binder, which was yet another organization task, but the daily task sheets provided structure for the day-to-day work on the projects.

Students used the daily task sheet to identify their task for the day and spell out what they needed and the steps they intended to complete in that class period. Before the end of the class, we gathered students back together a few minutes to reflect upon what they had accomplished that day, and most importantly, where they needed to start the next time they came back to the project. We found that if students left a note to themselves about what they needed to do next, it was a great on-ramp for them to get going more quickly and productively the next time they came to class. These "danglers" as we called them provided a starting point to remind themselves of what to do next.

PBL Daily Task Sheet: What is your goal for today?

♦ What did you intend to accomplish today?

♦ What did you accomplish today?

♦ What is your self-grade for today?

Level 4 - I completed all of my goal and more so my goal setting was excellent	Level 3+ - I completed nearly all of my goal so my goal setting was very good	Level 3 - I completed most of my goal so my goal setting was good	Level 2+ - I completed some of my goal so my goal setting was OK. I will use some of today's goal for my next lesson	Level 2 - I completed little of my goal so my goal setting needs work. I will use most of today's goal for my next lesson	Level 1 - I completed none of my goal. I will use this as my goal for my next lesson

♦ Where will you start on the project next class?

♦ What will you need to bring or prepare for next class?

FIGURE 10.5 Daily task sheets kept students on task during PBL

Fitting Into the Crowded School Day

Even with these student supports, PBL was challenging in the typical school setting. One constant source of frustration to me while conducting PBL at Hillside was that classes were only 55 minutes long, which was really only 45 minutes once all the attendance and administrative work was done. Just like the 6th grade class I observed and wrote about in the last chapter, frequent interruptions of announcements and external visitors were a constant at Hillside. There was very little focused time available for the project work. We would have loved to have a two-hour block period where students could have ample time to ramp up, get lots of work done, and still be able to reflect and document their work before moving on to the next class. Judging by the number of students still working by the time the bell rang, I think they would have liked a longer PBL class, too.

Ideally PBL would expand from TechEd to include content from across students' classes. Within the current school structure, it was hard to coordinate with other teaching staff to build cross-disciplinary projects. The principal set aside a bit of school-wide professional development time for interdisciplinary discussions, but the other teachers had already laid out their sequences and weren't inclined to reshuffling things to synchronize with project needs that were likely to come up. This would require school-wide planning at least a year in advance and a shared motivation among all teachers to do it. With the efforts Nova Scotia is taking in PBL, I am hopeful that this will rise as a priority in school planning, but it will come along with the many other initiatives and priorities, so we will wait and see.

So, what are you thinking?

1. How could the strategies we used for PBL at Hillside—such as danglers and daily task sheets—be adapted to support engagement and learning in your classrooms, workplace, or daily life?

Inclusive Assessment of Project-based Learning

Over the next few years at Hillside, TechEd teachers refined the student materials to help each student navigate their way through the PBL process. With care and attention, they were able to engage nearly every student in their inclusive classrooms. Students who struggled in other classes were thriving in TechEd. Students who were high achievers in their other classes were also doing well. PBL was an equalizer that let everyone play to their strength.

Reading and writing, however, were still difficult for many of these students, so Ms Bradbury was spending a lot of time finding different ways to help them express what they knew. Also in her previous classes, graded student work was often embedded in worksheets with drawings and ideas from which Ms Bradbury drew her term assessment, but at the end of the term, many students' worksheets were still missing from their binders or were incomplete. Ms Bradbury had no concrete evidence to represent their substantive progress towards the outcomes. For the final few weeks of each term, she had previously dedicated the first portion of each class to the process of reminding, coaching, and finally hand-holding students to get their completed worksheets in the binders. She often gave a lecture, repeatedly, that said: "I watch you all here in class. I know you are brilliant. You show me you are brilliant every day. But that is not what your binder says. Your binder says you are failing. I want you to get all your work in, so you and your parents and the world out there can see how brilliant you are." The grade was a measure of success, and the way to the grade was through the worksheets. It wasn't working for much of her class. This forced Ms Bradbury and me to think a lot about how to assess the process of PBL, rather than the product.

Measuring Progress

Ms Bradbury started letting students use their phones (or borrow one from us) to take photos of their work or use drawings, video, code, and conversation to explain what they had done. This was one layer of modification, avoiding some of the dependence on text, but it didn't get at the heart of the problem. We were still thinking about rubrics to evaluate final products. We were still missing out on all the progress and learning that took place in the process of PBL, especially for those learners who had trouble completing an entire project or whose final product didn't incorporate all they had learned along the way. Ms Bradbury created an "on-the-fly" assessment rubric that they could use to assess students' progress as she was supporting them on their projects. She copied a list of student names (one per row) and created a set of columns for each of the pertinent outcomes she was looking for (Figure 10.6). As she circulated through the class, when she observed any opportunities for the outcome to be demonstrated, she marked a grade (1–4) in the cell denoted by the student's name and outcome. Often she added a note on the side of the page to remind herself of the evidence for the mark.

Towards the end of the marking term, Ms Bradbury usually had several marks (from different days) from each student for most of the outcomes. She made sure to spend time with any students whom she had missed so she could evaluate each student one on one through questions about their project, going through their binder with them to provide further evidence. Evaluating the student at the same time as discussing the project gave each child a chance to show what they knew in their own way, in a natural setting of the work at hand. This gave Ms Bradbury a better understanding of students' efforts and talents than a summative, static rubric that only evaluated the final product. By using assessments that sought to reveal what learners can do, rather than focusing on deficits, we were able not only to provide each student with what they needed, we were also able to do it right at the moment they needed it and reveal their strengths—even to themselves. This, we found, was key to helping each learner succeed.

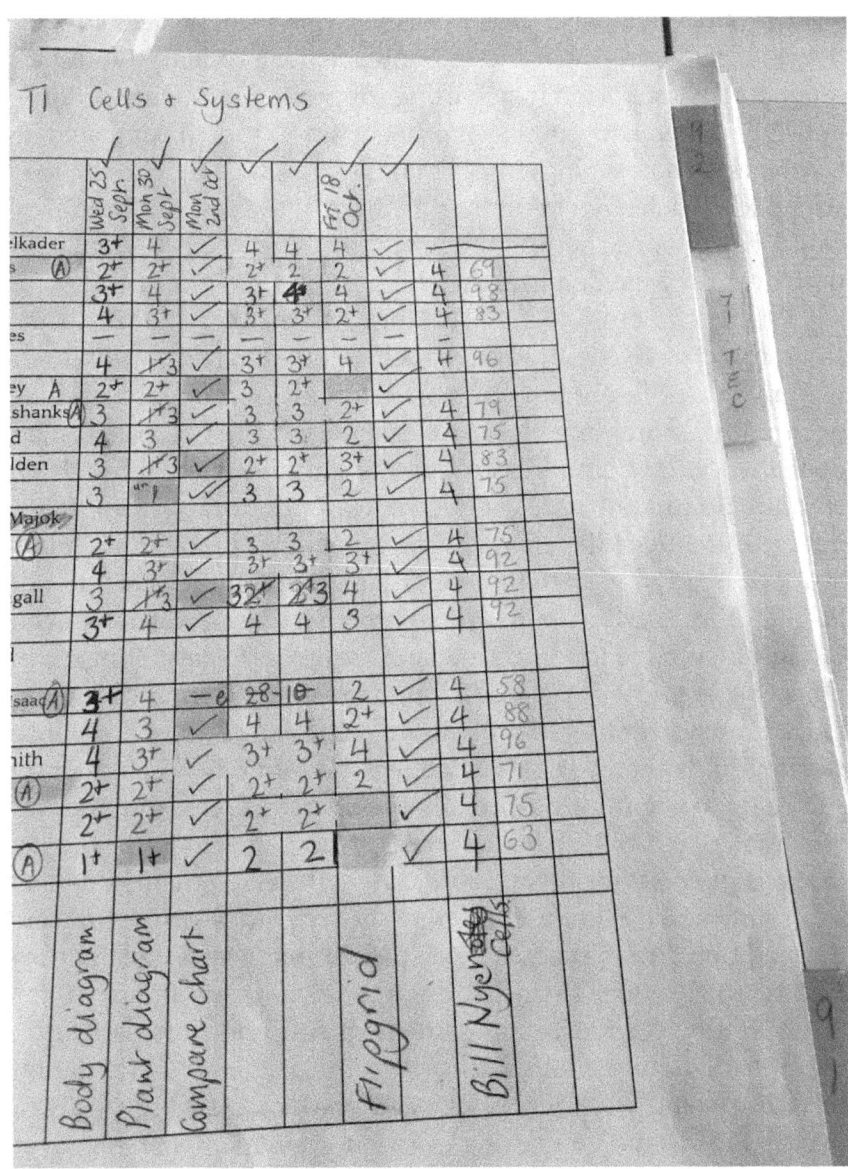

FIGURE 10.6 Completed process assessment for PBL
Source: Photograph by the author

Assessment in the Moment

Through this type of flexible, real-time, formative assessment, Ms Bradbury was also able to respond to student needs in the moment she saw them arise. Some of her students would sit quietly all class period just waiting for a ruler or a pencil, getting nothing done. They may have been too shy or disengaged to advocate for something simple that they needed to get to the next step of the task. By circulating through the class, Ms Bradbury could easily spot the idle student and hand them the ruler or whatever they needed to move on.

When students were working in groups, the real-time assessment gave Ms Bradbury a chance to sit with individuals who may need more time to work through an emotional, social, or cognitive stumbling block. At least once in every class I visited over the years, I saw Ms Bradbury help a student through an emotional situation. Most often, it was due to a lack of confidence. By squelching the echo of a negative voice in a child's head, or by giving them time to organize their thoughts without intimidation, or by reminding them that they are smart even though they have loud internal and external voices telling them otherwise—every day Ms Bradbury saved a kid from drowning in their own negativity. Motivation, self-esteem, and persistence can have a dangerous snowball effect. A child that is consistently subjected to negativity, especially about themselves, at home and at school, may have a hard time rising above that to find a project that has purpose for them. Struggling with assessments that feel irrelevant while others seem to get good grades easily, furthers the isolation and inadequacies a student may feel, which in turn often further erodes their identity, agency, and autonomy.

Real-time formative assessment during the process of PBL promotes learner-centered learning and enables differentiation of the learning experience for each student. As GLF posits, a key to PBL is the "value it places on student ownership over learning and the development of student expertise and agency as a resource for learning." With authentic assessment during the process of PBL, we were able to approach this goal.

This novel approach is also important because, as GLF also notes, PBL can provide a basis for open and honest conversations about educational equity. Many of the academics and practitioners they studied had been previously using models of instruction rooted in deficit views. Their methods involved testing and assigning levels to groups of students. When teachers had collegial conversations about PBL, it was seen to disrupt the use of the deficit-based approaches and supported the use of more equitable, asset-based practices of education. In other words, they worked to the strength of neurodiversity rather than trying to fix it.

So, what are you thinking?

1. How could the strategies we used for process-based assessments for PBL at Hillside be adapted to assess learning in your classrooms, workplace, or daily life?

Bibliography for Chapter 10

Asbell-Clarke, J. (2017, November). *We know more than we can tell.* TedxBeaconStreet.

Haladyna, T. M., & Downing, S. M. (2004). Construct-irrelevant variance in high-stakes testing. *Educational Measurement: Issues and Practice, 23*(1), 17–27.

Hira, A., & Hynes, M. M. (2018). People, means, and activities: A conceptual framework for realizing the educational potential of makerspaces. *Education Research International, 2018*(1), 1–10.

King, J. (2018). *3 Ways project-based learning can address equity* [Video]. Edutopia. https://www.edutopia.org/video/3-ways-project-based-learning-can-address-equity/

Lin, Q., Yin, Y., Tang, X., Hadad, R., & Zhai, X. (2020). Assessing learning in technology-rich maker activities: A systematic review of empirical research. *Computers & Education, 157*, 103944.

Lucas Education Research. (2021). *Rigorous project-based learning is a powerful lever for improving equity.* Lucas Education Research.

Mislevy, R. J., Almond, R. G., & Lukas, J. F. (2003). A brief introduction to evidence-centered design. *ETS Research Report Series, 2003*(1), i–29. https://www.lucasedresearch.org/wp-content/uploads/2021/08/Equity-Research-Brief.pdf

11

Strategies for Inclusion— Game-based Learning

Let Me Know When She Stops Talking

One of the annual conferences I have enjoyed most in my career is the Games+Learning+Society (GLS) conference that ran for about a dozen years at the University of Wisconsin in Madison. One of my favorite session types at GLS was the Hall of Failure where researchers could give postmortems of work gone awry. One year, Dr. Constance Steinkuehler, a pioneer in game-based learning, gave a wonderful Hall of Failure talk called, "Let Me Know When She Stops Talking." It was an account on some out-of-school workshops she and her team had run to engage urban youth who were chronically disengaged at school. Steinkuehler's team used the popular digital game, *World of Warcraft*, as a focus of the workshop, and they were hoping to hold real-world discussions with the youth as they were playing the game. Their research had already shown evidence of scientific inquiry practices among *World of Warcraft* players, and they hoped to use the game to motivate marginalized youth to engage in further STEM learning. The team had prepared activities and discussion prompts meant to facilitate connections between gameplay and scientific activity outside the game.

Once the workshops were going, the researchers noticed a boy in the back of the room who sat up to attention and was highly engaged in game-playing, but who slumped back in his chair and whipped up his hoodie over his head as soon as Steinkuehler stood up for the discussion. Steinkuehler reported she heard him lean over and say to his friend, "Let me know when she stops talking" as he disengaged. He was only interested in playing the game, not in her way of learning more about it.

Fortunately, this workshop had a serendipitous happy ending. When realizing that her discussion prompts and magnetic personality were not enough to keep these kids' attention, Steinkuehler's team presented a rack of books from the library that were specifically about improving one's game in *World of Warcraft*. Though most of the kids in the workshop were reading well below grade level in school, and these books were more advanced, many of them figured out how to read what the books were saying because they needed the information to do what they wanted to do, which was beat the next level of *World of Warcraft*.

The game was a hook.

What Is Game-based Learning?

In 2009, my colleagues and I founded the Educational Gaming Environments group (EdGE) at TERC to study game-based learning in STEM. We saw digital games as a new frontier for STEM problem-solving education. It was a place where many learners chose to spend a lot of time, and also provide a rich environment to employ many of the learner-centered, constructivist learning opportunities that learning science research suggested. A friend and mentor of mine, Dr. Richard Halverson, once said to a roomful of learning scientists, "It's like the game designers read the learning science literature, and they have implemented it a heck of a lot better than we have."

The top-rung video game companies, those that produce high-budget, high-profile games, have motivation and engagement down to a science. The business model for these companies relies upon making players want to persist in the game, so developers make "sticky" games—games that keep people in flow. A good game teaches the skills and understandings required for success as part of the fun of the game. Otherwise, most people wouldn't play them.

What *Portal* Taught Me

I, myself, am not a big gamer. I hadn't played many video games at all back when our team started studying game-based learning. As I delved into the field, however, I quickly found a game called *Portal*. *Portal* is set in an environment with imaginary laws of physics. The series of game puzzles requires understanding and finding increasingly complex ways to deal with the new physics. At the beginning of each puzzle, I never had a clue as to what I was meant to do to solve the puzzle. Nothing was apparent at first, but after ample experimentation through trial and error, by the time I solved each puzzle, I could tell you exactly what I had learned to solve it. I was hooked on the power of game-based learning.

Not only is *Portal* a super fun game, it also comes with a developer commentary that one can follow along with the game. In this commentary, the developers talk about all the play testing they did with users in the development of the game to keep them in that sweet spot between frustration and boredom (similar to Vygotsky's notion of a ZPD or Csikszentmihalyi's concept of flow) and how their observations helped them hone the mechanics to keep every player at peak engagement. It was a matter of figuring out the pain points. Where did people get lost? And what kind of slight boost was just what they needed to get to the next step without giving too much away? As game designer and futurist Jane McGonigal explains:

> In a good computer or video game you're always playing on the very edge of your skill level, always on the brink of falling off. When you do fall off, you feel the urge to climb back on. That's because there is virtually nothing as engaging as this state of working at the very limits of your ability.

Learning in Commercial Games

In 2012, the MacArthur Foundation, the Bill & Melinda Gates Foundation, Entertainment Software Association, and Electronic Arts joined together in an effort called GlassLab to bring this kind of stickiness of top video games to K–12 classrooms. Electronic Arts (EA), a top game company, teamed up with experts in learning and assessment models to create a few high-powered learning games. They built an educational version of Sims games called *SimCityEDU*. They also built a game called *Mars Generation One: Argubot*, designed to measure scientific argumentation among middle schoolers. Unfortunately, GlassLab did not create the revolution they imagined for education. After reaching over a thousand teachers in six years, they were not able to build a financially sustainable model to continue the work.

While not creating the commercial impact they hoped for, these early game-based learning efforts helped to steer researchers, including my team at TERC, towards innovative models for the support and assessment of STEM problem solving. The design of game-based learning assessments has helped move researchers away from a traditional assessment design of formal pre/posttests, which may provide barriers for neurodiversity. Instead, stealth assessments in games measure learning using tasks embedded within the gameplay itself to "support learning, maintain flow, and remove (or seriously reduce) test anxiety, while not sacrificing validity and reliability."

Implicit Game-based Learning

As philosopher and scientist Michael Polanyi said, "We know more than we can tell." There is a layer of implicit learning that occurs outside of formal or school-based learning. Early examples of implicit learning include the mathematical abilities of gamblers calculating odds at the racetrack, Brazilian street children using early algebra skills in their vending of fruit and snacks, and housewives calculating "best buys" at the supermarket. Today, there are millions of neurodivergent learners exhibiting their extraordinary talents in video games.

Many learners have rich implicit knowledge—understandings and abilities that they may be unable to articulate explicitly. As players repeatedly try, fail, revise, and retry to solve or beat a game under increasingly complex situations, they may demonstrate patterns of behaviors in the game that are consistent with implicit understanding of underlying game content, even if they don't talk about it. In games specifically designed for learning and assessment, Jan Plass and colleagues suggest a strong alignment between the following components: (a) the game mechanic—what a player does to succeed in the game; (b) the learning mechanic—what the designers intend that the player learns in the game; and (c) the assessment mechanic—the evidence of players' learning that can be gleaned from their game activity. When games are designed with these facets aligned, analysis of players' activity within the game may demonstrate meaningful evidence of implicit learning.

Tapping into this implicit game-based learning as a form of learning assessment may be particularly critical for the support of neurodivergent learners. Many neurodivergent learners love games, and many neurodivergent learners excel in games. In fact, my colleagues at Fenworks run a high school e-sports competition with teams drawing from across North Dakota and Minnesota, and the winning team of this year was from a school for autistic learners. But more importantly, games may provide a window into neurodivergent learning, learning that may go otherwise unseen.

So, what are you thinking?

1. How do video games fit into your classroom, workplace, or daily life?
2. Do you see games as a useful way to connect with learners?
3. What do you think may be potential detriments to game-based learning?

Martian Boneyards

The first game our team designed was a science mystery game in a virtual world. Funded by the US National Science Foundation's

Strategies for Inclusion—Game-based Learning ◆ 143

FIGURE 11.1 In a virtual mystery game called *Martian Boneyards*, players engaged in four months of scientific problem solving to solve the mystery

Informal Science Education program in 2009, we partnered with top-end designers in the high-definition virtual world of Blue Mars. In Blue Mars, avatar representations of the players could walk and run, use a chat window, and use embedded tools that were activated by clicking on a heads-up display. We created an abandoned science center and extensive grounds around the center (see Figure 11.1) in Blue Mars to host a prototype science mystery game we called *Martian Boneyards*. The science center was surrounded by beautiful grounds and a gruesome mystery. We had placed (virtual) bones from humans, including Neanderthals, and chimpanzees as well as other animals in the science center surroundings in a manner that could seed a variety of evidence-based storylines.

Three members of our design team (including me) played characters in the game. We created avatars and interacted with other players as if we were one of the Blue Mars explorers who had discovered the site. We asked the members of the Blue Mars community to help us figure out what had happened, and how all the mysterious bones got there. Players were encouraged to take the story in any direction for which the community agreed there was ample evidence. For example, when they discovered a skull lying some distance away from a partial skeleton located near a stream that had lemur bones nearby, they asked the questions: Was it moved from its original location by scavengers or water or other means? Did a skeleton wind up at the base of a cliff due to being pushed or having fallen, or just dying in that place? These types of questions and the subsequent scientific topics of inquiry were left open for the players to investigate (or not).

Measuring Scientific Inquiry in *Martian Boneyards*

As the game played out over four months, we used the backend digital data logs generated by players' activity in the game to measure the extent of scientific inquiry players conducted in the game. We designed the game mechanics of the game to align with three phases of scientific inquiry: data gathering, analysis, and evidence-based theory building. Each player's click on any tool was associated with one of the three phases of inquiry, so we counted the clicks for each player and each tool. To accompany this simple quantification of the extent of activity with the inquiry tools, we conducted participant observations as we interacted with the players in the context of the game.

After the game was over, we also conducted a review of the scientific content in players' postings for a larger picture of the scientific inquiry in the game. The quality of the scientific content generated in *Martian Boneyards* was measured through a review of player-generated artifacts by a panel of three independent scientists from related fields. The reviewers judged that the player community engaged in sustained scientific inquiry—questions, making claims, substantiating claims with evidence—to an extent that would be considered very good (rating of 4 out of 5) in an undergraduate introductory science course. One reviewer commented, "Those top players reminded me of those students you get once in a while that just have a burning desire to learn."

We found that although the typical *Martian Boneyards* player was a 36-year-old white male, female players conducted more than 66% of the scientific inquiry in the game, especially in the advanced analysis and theory-building phases. This was particularly interesting when considering the underrepresentation of women in STEM. The two most active players in the game were both female. One top player, KalW, was a 52-year-old bus driver who never went to college. When we interviewed her after the game was over, she said she enjoyed learning the science in *Martian Boneyards* and talking with her husband (who did not play the game) about the questions she had. When asked what compelled her to spend over 150 hours in the game, she explained that she loved *Martian Boneyards* for the layout, the idea, and the learning. She and others referred to their need as gamers to solve the mystery of *Martian Boneyards*.

The player with the highest level of activity, Jespau, spent over 200 hours in the game, and said she spent many more days outside the game collecting information and organizing files for *Martian Boneyards*. In a postgame interview, she explained that she spent so much time playing the game because:

> Martian Boneyards allows me to learn and play. . . . I have a folder with over 200 things in it! . . . It was a matter of getting to the conclusion by whatever means. If it means I have to learn science, then that's fine . . . I am a gamer. We never give up!

Jespau told us she was an agoraphobe and hadn't left her apartment in months except for absolute necessities, and that is why she loved being in a virtual world. Jespau had avidly gathered information and worked to solve the mystery, but she often minimalized her own work saying, "Oh, these are just all my notes—I don't know why anyone else would want to read them." Later in the game, her expertise was recognized by other players, and they started calling her "Doc" and going to her for updates on the mystery findings. Ultimately, the community chose her for the top scientist award in a final ceremony celebrating the solving of the *Martian Boneyards* mystery. She said it was one of the greatest honors she'd ever experienced in her life.

The *Martian Boneyards* experience taught our team several important takeaway lessons. The first was that when placing the learner at the center of the game experience, and by allowing nimble design and facilitation, we were able to engage some traditionally marginalized nonscientists in authentic scientific inquiry. We also learned that the digital records generated through STEM problem-solving games could provide rich fodder for novel and more inclusive assessments of STEM learning. Finally, we learned that games showed particular promise for reaching marginalized learners.

Leveling Up

Having proven to ourselves that we could use simple analytics to measure the extent of scientific inquiry in *Martian Boneyards*, our team wanted to extend these methods to measuring STEM conceptual knowledge and skills for a broader audience. Our aim was not to create more gamers, it was to use the already popular activity of digital games to broaden participation in STEM, and to learn how to better support and measure STEM learning from marginalized audiences. This work piqued the interest of the US National Science Foundation and it generously provided more funding.

A few of the grants we received were specifically for high-risk innovations to help change the way educators think about assessments. For example, we were funded to design a set of games to support and measure game-based learning in high school science classes. We were excited by the kind of measurement we could do with digital logs in *Martian Boneyards*, but virtual worlds were still fairly exclusive to white, affluent, male tech enthusiasts at the time. We wanted to reach a broader audience. We designed a set of tablet games we called Leveling Up games that used popular game mechanics to build implicit learning of high school science content and would play on common platforms that were proliferating through schools. *Angry Birds* was very big at the time, and although we used different phenomena, we wanted a similar repetitive game that honed in on a simple physics concept.

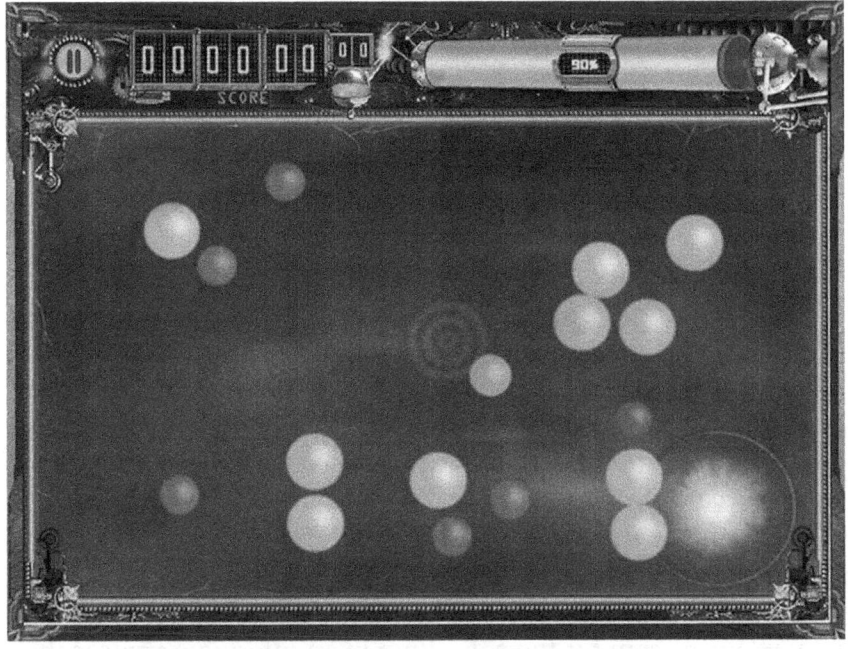

FIGURE 11.2 A tablet game called *Impulse* helped teach high school students about Newton's laws of motion

Building Game-based Learning Assessments

Game-based learning assessments in research often use an evidence-centered design (ECD) model to establish a logically coherent and evidence-based argument describing how the assessment tasks in the game are aligned with the claims being made about the game-based learning. Grounding in an ECD model guides designers and researchers to: (a) define the claims to be made about learners' competencies, (b) establish valid evidence of the claims in gameplay behaviors, and (c) determine the nature and form of gameplay activities that will elicit that evidence. This design is very useful to parse out intended learning outcomes but tends to front-load a lot of the decisions about what game-based learning will occur to early phases of the design process. In other words, the designers have to know what to expect from players before they build the game. Our team has been using a more emergent ECD approach. We design a game that enables learners to dwell in a rich learning environment that reinforces the learning mechanic, and we analyze players' behaviors as they grapple and succeed in the game to decide what will constitute evidence of learning based on a theoretical framework. This approach leaves us more open to revealing evidence of learning from a wide range of learners that we may not have predicted beforehand.

When determining the evidence of learning, we use highly systematic and rigorous methods to avoid the common designers' fallacy of measuring only what we expect or want to see. We typically use two primary sources of data. The first is extensive video recordings of the players' faces, voices, and screen capture as they play the game. This allows us to see their expressions along with their gameplay and hear comments, which are particularly useful if they are playing and discussing the game with another person. We encourage participants to play in pairs or we may prompt them for their thinking in an interview style as they play the game to ensure they verbalize as much as possible their thoughts about why they are doing what they are doing in the game. Concurrently, we collect digital data logs of all consenting players' "click" activity in the games. We record every "click," by which we mean every potentially salient action in the game, along with information about the game settings at the time, and we tie that information to a player ID and a timestamp. These data logs provide a digital footprint of each player's activity in the game.

We analyze the video recordings of gameplay by having teams of researchers who are deeply familiar with the theoretical framework for the learning game and the STEM content, but who are not the original designers of the game. This allows some objectivity to the analysis. The video coders label the players' actions and strategies and come to an agreement on what constitutes evidence of learning. The researchers isolate the salient strategies and actions that are important to learning in the game and break those strategies down in to individual click patterns that constitute evidence of each potential learning outcome. Once we have valid and reliable patterns showing evidence of learning, we then build data-mining models to detect these patterns automatically in large amounts of data logs for all players. Sometimes we use automated detection models that can train themselves, so that as we add more data from more players the models improve in reliability and predictability.

While this type of automated game-based learning assessment is still a resource-intensive research endeavor needing much more design and validation work at large scale, it serves as an important model of learning and assessment as we move to inclusive education. These types of emergent data-driven assessments may be critical in our attempt to reveal the STEM talents of neurodivergent learners. We can use the information from these game-based assessment models to inform the learner on their own progress in the game. Gamers are accustomed to dashboards that show their progress or "powers" in a game. This provides a fun and potentially less intimidated way to provide feedback to students that may not come with the stigma of grades and traditional assessments. It may be a novel way to provide formative assessment to learners so that they can help themselves become better learners.

Adaptive Games

Potentially even more powerful for neurodivergent learners, the information from game-based learning assessments can also be used to feed back into the gameplay itself. This is called *adaptability*. As described earlier, many professional game designers use the players' progress through the game to adapt the pace or complexity of the game so that players stay in their sweet spot of fun and engagement. With data-mining models, this can be done in real time and can be used to optimize learning. Some simple learning technologies can detect where learners are struggling, but they keep providing similar content, trying to move the learner towards mastery through repetition. This is not always effective and can lead to frustration and alienation without careful scaffolding and breaks. Imagine being thrown over and over into the same obscure puzzle you have no idea how to solve. Some game-based educational data-mining models, however, are sophisticated enough to detect where the player is going wrong and make useful suggestions, or even detect boredom, frustration, and distraction so that the game can be adjusted to accommodate for those.

Our team at EdGE at TERC is looking ahead to future work where we can build adaptive models with the scaffolds we developed to support executive function in games like *Zoombinis* as described in Chapter 12. We envision a game that can sense distraction and re-guide a learner's attention to salient information, or detect when a learner is losing track of information and provide a working memory scaffold. This would be the kind of game that could truly differentiate and support neurodiversity in STEM education. Data-mining models could determine when and how to fade the scaffolding to help each learner reach autonomy, creating a ZPD for each player. This, of course, is a complex challenge and will take extensive and careful research.

Bridging Implicit Game-based Learning to Explicit Classroom Learning

In the meantime, one of the most important uses for the information revealed through data-mining models and game-based assessments could be to provide educators with necessary information for differentiated teaching and learning. During our implementation studies of the Leveling Up games, we studied how teachers implemented the games in their classroom and how we could help them leverage game-based implicit learning for improved high school science class learning. We saw that a powerful strategy for game-based learning includes bridging—where teachers provide activities and strategies to connect the implicit knowledge to explicit classroom outcomes.

Our research on Leveling Up showed us that when students' gameplay logs showed evidence of increased use of strategies consistent with understanding of Newton's laws of motion, and when their teachers used ample bridging to physics learning in the classroom, those students were able to show greater growth on pre/posttests measuring their understanding of similar physics content. In other words, when teachers build on the "aha" moments that students have in games to make connections with the content being covered in the classroom, their students learn more.

There are two large obstacles in bringing game-based learning assessment to scale for effective bridging and differentiation by teachers. First of all, the amount of work to get reliable and validated detectors for each game is extensive, and the work is quite specific to each game. Secondly, it is very difficult to design an interface for teachers (often called a dashboard) that presents enough information to be meaningful for a teacher without it being so much that it overwhelms them in the middle of class. This is yet another holy grail of educational designers.

Game-based learning has opened doors for researchers, designers, and educators to new ways of engaging a broader audience of learners in STEM problem solving, and can also lead to innovative assessment of that STEM problem solving. Pedagogical avenues such as game-based learning, when done well, can reform who thinks of themselves as a STEM problem solver and who becomes a STEM problem solver. These are the types of pedagogical reforms that can help neurodiversity thrive.

Bibliography for Chapter 11

Asbell-Clarke, J., Edwards, T., Rowe, E., Larsen, J., Sylvan, E., & Hewitt, J. (2012). Martian Boneyards: Scientific inquiry in an MMO game. *International Journal of Game-Based Learning (IJGBL)*, 2(1), 52–76.

Asbell-Clarke, J., Rowe, E., Bardar, E., & Edwards, T. (2020). The importance of teacher bridging in game-based learning classrooms. In *Global perspectives on gameful and playful teaching and learning* (pp. 211–239). IGI Global.

Collins, H. (2019). *Tacit and explicit knowledge*. University of Chicago Press.

Farber, M. (2021). *Gaming SEL*. Peter Lang Verlag. Retrieved May 2, 2023, from https://doi.org/10.3726/b18044

Hamari, J., Shernoff, D. J., Rowe, E., Coller, B., Asbell-Clarke, J., & Edwards, T. (2016). Challenging games help students learn: An empirical study on engagement, flow and immersion in game-based learning. *Computers in Human Behavior*, 54, 170–179.

McGonigal, J. (2011). *Reality is broken: Why games make us better and how they can change the world*. Penguin Random House.

Mislevy, R. J., Almond, R. G., & Lukas, J. F. (2003). A brief introduction to evidence-centered design. *ETS Research Report Series*, 2003(1), i–29.

Nunes, T., Schliemann, A. D., & Carraher, D. W. (1993). *Street mathematics and school mathematics*. Cambridge University Press.

Plass, J. L., Homer, B. D., & Kinzer, C. K. (2015). Foundations of game-based learning. *Educational Psychologist, 50*(4), 258–283.

Qian, M., & Clark, K. R. (2016). Game-based learning and 21st century skills: A review of recent research. *Computers in Human Behavior, 63*, 50–58.

Rowe, E., Asbell-Clarke, J., & Baker, R. S. (2015). Serious games analytics to measure implicit science learning. In C. S. Loh, Y. Sheng, & D. Ifenthaler (Eds.), *Serious game analytics: Methodologies for performance measurement, assessment, and improvement* (pp. 343–360). Springer International Publishing.

Rowe, E., Asbell-Clarke, J., Bardar, E., Almeda, M. V., Baker, R. S., Scruggs, R., & Gasca, S. (2020). Advancing research in game-based learning assessment: Tools and methods for measuring implicit learning. In *Advancing educational research with emerging technology* (pp. 99–123). IGI Global.

Shaw, R., Grayson, A., & Lewis, V. (2005). Inhibition, ADHD, and computer games: The inhibitory performance of children with ADHD on computerized tasks and games. *Journal of Attention Disorders, 8*(4), 160–168.

Shute, V. J., & Kim, Y. J. (2014). Formative and stealth assessment. In *Handbook of research on educational communications and technology* (pp. 311–321). Springer International Publishing.

Shute, V. J., Ventura, M., & Ke, F. (2015). The power of play: The effects of Portal 2 and Lumosity on cognitive and noncognitive skills. *Computers & Education, 80*, 58–67.

Steinkuehler, C., & Duncan, S. (2008). Scientific habits of mind in virtual worlds. *Journal of Science Education and Technology, 17*, 530–543.

Steinkuehler, C., King, E., Alagoz, E., Anton, G., Chu, S., Elmergreen, J., Fahser-Herro, D., Harris, S., Martin, C., Ochsner, A., Oh, Y., Owen, V. E., Simkins, D., Williams, C., & Zhang, B. (2011, June). Let me know when she stops talking: Using games for learning without colonizing play. In *Proceedings of the 7th international conference on games+ learning+ society conference* (pp. 210–220).

12

Strategies for Inclusion—Computational Thinking

Renee—Creating a World of Cats

One of the grade 9 Tech Ed students in Mr. Jerrett's class, whom I'll call Renee, had autism. The first thing I noticed about Renee was that she wore loose clothes made of soft fabric and she appeared more juvenile than the other girls' styles. Her hair was tangled in knots. Her social behaviors were awkward, and she was alienated from the rest of her class.

Renee seemed to live in her own world. While she was withdrawn and rather sullen much of the time, she could also be highly excitable when talking about the thing that interested her—cats. Once Renee learned I worked with game designers, she marched right up to me and wanted to tell me all about the game she wanted to build. She set off into a rapid-fire monologue about a game with a cast of many cats and all their antics. It was very difficult for me to understand her. Most of her words were mumbled or seemed nonsensical to me, but she clearly had a vision in her head, and she was clearly very amused by this vision.

We introduced Renee to Scratch, an introductory coding environment that was designed by Mitch Resnick and his colleagues at MIT to enable learners to build animations and games while learning the basics of computer programming. I helped Renee create an account and, within minutes, she told me to go away.

She had found everything she needed on the Scratch site, and therefore no longer required, nor desired, any help from others.

Over the next two months, while the rest of the class worked on a variety of projects, Renee persevered diligently on her game. Occasionally, a burst of laughter would bubble up from her as she drilled persistently down on the keyboard. A few times when I'd stop by her workstation to see how she was progressing, she'd show me the latest clip of her cat's extensive dialogue or the brilliant green outlandish outfit she'd created for the cat with the leading role. I wasn't sure what type of product Renee was creating, or what the storyline was, but she sure was keeping busy.

Because Renee was on an IPP, she had modified outcome expectations, and what she was doing with Scratch would be acceptable for a passing grade, especially in TechEd, so we were not worried. We figured we could spend our time with other kids who demanded more attention. At the end of the term, Renee signed up for a slot to demonstrate her project to the class. Neither Ms Bradbury nor I knew what to expect, but I will be honest and say my expectations were not high. I was a bit anxious about what might happen when she stood in front of the class spewing her erratic tales of cats that no one could understand. Kids can be cruel, and I was afraid we were setting her up for a disaster.

Renee gave a 20-minute presentation in which she did not stop her rapid-fire monolog for one moment. No one else in the class could understand Renee's story. The narrative meant nothing to us. We couldn't make sense of what she was saying. But what was clearly appreciable by everyone in the class, however, was the extraordinary amount of intricate coding and artwork in her animation. Renee had mastered complex scene changes, the development of a long list of characters each with their own motion and prescribed dialogue, and built artistic transitions, all using sophisticated features in the Scratch programming environment. Renee had built her own functions in Scratch that she named to correspond with each character, giving them their own ways of walking, jumping, and dancing around the screen. She had created time delays and event-driven phenomena to add a surprise element to her animation. As far as grade 8 coding projects I've seen in my career, it was a masterpiece.

Scratch gave Renee a way to communicate her zany, cat-filled world to us. The action of putting blocks of code together in patterns and algorithms that enabled her to make her world come alive delighted Renee in a way her teachers, her parents, and her classmates had never seen before. Scratch gave Renee her voice.

Computational Thinking

One way to be more inclusive of neurodivergent learners in education is to choose areas that are aligned with common cognitive strength, rather than focusing on deficits. An interesting disciplinary area that is emerging in K–12 education and has substantial overlap with common cognitive talents associated with neurodiversity is *computational thinking*.

Computational thinking is relatively new in education, and thus does not yet have a widely accepted definition. Computational thinking focuses on the foundational practices required to work with computerized systems. Originally thought of as a conceptual basis for computer programming, computational thinking is now also recognized as an important form of generalized problem solving focused on efficiency and reusability. Computational thinking practices include *problem decomposition*, which is decomposing complex tasks into simpler, more manageable tasks. Computational thinking also includes *pattern recognition* and *abstraction*, which involve seeing commonalities in problems and their solutions, and generalizing those solutions so that one can develop *algorithmic thinking*, which enables repeatable and reusable problem-solving techniques. These foundational practices of computational thinking are not only used in programming and STEM innovation, but they are also fundamental to everyday problem solving (e.g., making a meal or cleaning up the classroom) as well as many school-based learning activities (e.g., solving a math problem, conducting a science experiment, or writing an essay).

Computational thinking practices also present an interesting intersection with the extraordinary talents of many neurodivergent learners. Autism researcher Dr. Andrew Begel says autistic people often excel at sifting through information. He suggests this may be because they must spend much of their lives trying to find salient signals through all the background noise associated with the sensory overload they experience, so they get really good at filtering through it. Begel notes that since all that noise and stimuli are bombarding them all the time, they need to hone really good mechanisms to sift and sort out the pertinent signals, and that comes in handy as a programmer. This process of sifting through information is not only a talent, but it is also something several autistic employees with whom I spoke are keenly interested in doing.

Similarly, abstraction—generalizing solutions to a broader set of problems—may come naturally to learners with ADHD or dyslexia because their brains may make connections across problems more easily. While systematic thinking is useful in algorithm design, so is creativity and thinking outside the box. In grades 3–8 classrooms where my team has been studying computational thinking, teachers often tell us their students who struggle in other areas are the first to solve computational thinking puzzles. The formerly struggling student becomes the class rock star, the "go-to" resource for computational thinking. This can be transformative as other classmates, and the learners themselves, begin to see themselves as competent and clever.

Unfortunately, many neurodivergent learners do not have an adequate opportunity to engage in new disciplines such as computational thinking since much of their education may be devoted to remediation and/or with therapists, speech pathologists, and other specialists. For this reason, some STEM education efforts are providing opportunities specifically designed for neurodivergent learners. For example, Microsoft has developed coding camps for first-year college students with autism to help strengthen the collaboration and communication skills they'd need to succeed in the computer science workforce. They found students gained improved programming skills, increased confidence in communication, and better experiences working with others in these camps, and that students valued the opportunity to practice expressing ideas to their peers and working out differences of opinion with their teammates in a safe space such as a camp.

Computational thinking allows many neurodivergent learners to highlight their strengths as opposed to try to fix what's broken. These generalized problem-solving practices also provide learners with skills to deal with a rapidly advancing and increasingly complex technology-laden society. We need computational thinking in our future, and we already have many in the waiting if we choose to embrace their talents.

Zoombinis

In the past decade, with generous support from the US National Science Foundation, our team at EdGE at TERC has been studying the overlap between neurodiversity and computational thinking in US schoolchildren in grades 3–8. We started this line of research by studying the learners' development of computational thinking practices while playing a logic-puzzle game called *Zoombinis*. *Zoombinis* is a computer game created at TERC in the 1990s by my colleagues, Scot Osterweil and Chris Hancock, who wanted to engage learners in a fun way to think about data and databases using fun logic puzzles. The game sold over a million copies and won several awards from educators and parents' groups. In 2015, TERC re-released *Zoombinis* for tablets and a browser-based version that can run on many common devices.

Zoombinis consists of a series of 12 puzzles, each with four levels of complexity, in which players are charged with transporting packs of zoombini characters (16 at a time) through perilous puzzles to the ultimate safety of a land called Zoombiniville. Each zoombini has one of five different types of hair, eyes, nose, and feet. Players use these combinations of attributes to send the zoombinis through the sorting, matching, and sequencing challenges of each puzzle. The attributes of the zoombinis, and associated rules of each puzzle, change with each new round of play so players must find methods of solving the problems, not just "the right answer." Other puzzles in the game apply similar logic and CT practices in different contexts, such as identifying the exact combination of pizza toppings to satisfy hungry, but picky, trolls.

Zoombinis Puzzle 1: Allergic Cliffs

In the first *Zoombinis* puzzle, called Allergic Cliffs, stone cliffs are "allergic" to certain values of zoombinis' attributes (hair, eyes, nose color, or feet type). Players are challenged to figure out which zoombinis can cross which bridge without being sneezed back. They can use trial and error, but they have a limited number of sneeze-backs before the bridges collapse. In the example shown in Figure 12.1, the bottom cliff is allergic to flattop hair and the top cliff is allergic to all other hair types. This puzzle requires learners to decompose the problem of identifying one salient characteristic out of all the possible permutations. Players must develop systematic methods to test the zoombinis efficiently and recognize the pattern. When they abstract that pattern to a general

FIGURE 12.1 In the Allergic Cliffs puzzle in Zoombinis, one cliff is allergic to one value of one attribute and players must get their zoombini pack across the bridges without getting sneezed off

solution (the cliffs are sorting by hair), they can then develop an algorithm to solve the problem. This contrasts with players who stay in trial and error by randomly selecting zoombinis for each attempt.

Zoombinis Puzzle 2: Pizza Pass

Another puzzle in *Zoombinis*, called Pizza Pass, challenges learners to conduct similar systematic testing to determine the exact combination of pizza and sundae toppings that will satisfy each of the picky trolls (Figure 12.2). Players try different combinations of toppings and the trolls reject pizzas with at least one topping they don't like (by throwing them in the pit in front) and accept pizzas that have some toppings they like but not all of them (by throwing the pizza on the rocks behind them and asking for more toppings). To solve this puzzle, players often adopt systematic strategies such as trying one topping at a time or accumulating the toppings a troll has accepted until they get

FIGURE 12.2 In the Pizza Pass puzzle in *Zoombinis*, each troll wants a unique combination of pizza and sundae toppings and won't accept anything else

the right pizza and sundae. Because the puzzle changes each time it is played (and there are increasing levels of complexity for each puzzle), players must focus on developing methods (algorithms) for solving the problems as they progress. To do this, they decompose the problem into its components, look for patterns in what is successful (or unsuccessful), and abstract those solutions into generalized algorithms.

To support classroom teaching and learning of computational thinking with *Zoombinis*, our team designed a suite of supplementary activities that teachers in grades 3–8 could use to bridge computational thinking practices players experience in *Zoombinis* gameplay with problem solving in the classroom. One such activity was a physical recreation of the Allergic Cliffs puzzle in the classroom, where some students enact the rules underlying the puzzle (e.g., which attributes are rejected by which bridge), and other students are the zoombinis who are attempting to cross. This type of bridging activity (no pun intended) allows collaboration and sharing of practices and information that may not take place in online gameplay and may support some students who need help. Bridging activities also encouraged conversations in class that allowed the teacher and students to become very explicit with computational thinking, naming the problem decomposition, pattern recognition, abstraction, and algorithm design as they see it. Other bridging activities included activities that connected students' computational thinking practices in *Zoombinis* to other contexts such as collecting data in a science experiment, solving a math problem, and other puzzle games (e.g., *Mastermind*, *Guess Who?*).

Our research team also built a set of data-mining detectors and machine learning algorithms to identify computational thinking practices within learners' gameplay in grades 3–8. We were able to show that learners who exhibited more frequent use of computational thinking practices in the game also demonstrated greater improvement in external assessments of computational thinking. Teachers we worked with in related research reported that their neurodivergent students became class leaders in computational thinking activities, solving computational thinking puzzles quickly and teaching classmates new strategies. In the same study, special education teachers recognized an overlap between computational thinking and the executive function skills they were emphasizing for students' daily lives, such as breaking down problems into manageable tasks and recognizing patterns of behavior. We believe in particular that the metacognitive aspects of computational thinking, the need to plan and be explicit when programming a computer, may be a useful tool for neurodivergent learners.

Miaka—Designing Executive Function Supports

Miaka was a quiet grade 8 student who preferred to go by they/them pronouns. Miaka was quiet in class, a bit withdrawn socially, but was typically a good student. They could become very involved with a new lesson, often reading more than was assigned on a new topic, but they also often lost their homework sheet before it made its way into a binder or forgot to hand in an assignment. When I introduced *Zoombinis* to Ms Bradbury's class, Miaka was the first to soar through the entry-level puzzles and was already into advanced levels before I had gotten around to their desk for a check-in. Meanwhile, some of their classmates were still struggling on the first puzzle.

During the class, we occasionally heard exclamations of "I did it!" from Miaka or a "Yes!" with a discreet fist pump under their desk. We weren't the only ones that noticed. The girl who shared a double desk with Miaka, named Amarjot, was new to Canada and didn't speak much English. Amarjot kept peering over at the Miaka's computer screen, trying to see what was provoking such enthusiastic success. Slowly, Amarjot picked up tips by watching over Miaka's shoulder, and she too began to succeed in the game.

Once this pair of (previously ostracized) students was reaching points in the game that others hadn't yet encountered, the rest of the class took interest. They wanted to know what the pair was doing to be so successful. Ms Bradbury took a chance and asked Miaka to connect their iPad to the overhead projector, and asked Amarjot to come up front and help Miaka show the rest of the class what they were doing. With four hands, one iPad, an overhead projector, and very few words, the pair expertly showed the class how they solved the first three puzzles.

We amply celebrated the success of Miaka and Amarjot, allowing them to host the zoombini plush toy I'd brought into class on their desk for the rest of the day. The next time we brought out the iPads for *Zoombinis*, several classmates asked if they could sit near Miaka and Amarjot so they could ask questions if they needed help. The two kids no one had ever paid attention to before had become the class rock stars.

As the *Zoombinis* puzzles grew more complex, Amarjot sailed ahead. She had no problems keeping track of all the different pieces of information and putting it all together logically to find solutions to even the trickiest tasks. Miaka, however, grew frustrated because while they understood what they had to do to solve the puzzle and they could come up with clever strategies to do so, they were often losing track of a salient piece of information or forgetting something they had figured out earlier. They needed a bookkeeping tool to keep track of information.

As a designer, this was a golden opportunity for me. I had an enthusiastic, bright learner who knew how to do what they wanted to do but was facing an extraneous obstacle that could be overcome. I had also built a pretty solid relationship with Miaka by this time, so I was able to ask them for help in figuring out what they needed. I asked them to draw the tool they wanted to help complete the puzzle. They drew an illustration of a bookkeeping ledger with exactly how they wanted the information laid out.

For Allergic Cliffs, Miaka highlighted all the zoombinis on the screen with a specific attribute (e.g., pink shoes) so that they could keep track of them all while they were trying to solve the puzzle. For Pizza Pass, Miaka sketched out a data table that helped them record and organize which pizza and sundae toppings they had already tried and the associated results. With these tools—a highlighter to direct attention to the salient information and a graphic organizer that helped support working memory by keeping all the information visible and organized—Miaka would have no issues problem solving the rest of the puzzles.

INFACT

Based on the design input from Miaka and others, my team designed a set of digital scaffolds that we added to a few of the *Zoombinis* game puzzles. We used these tools as part of a bigger study called Including Neurodiversity in Foundational and Applied Computational Thinking (INFACT), a project funded by the Education Innovation and Research program of the US Department of Education. Our team at TERC has partnered with cognitive psychologists, STEM educators, special educators, and technology design researchers to build INFACT, a differentiated teaching and learning program for computational thinking in grades 3–8. INFACT includes games like *Zoombinis*, along with teacher bridging materials to integrate the game-based learning with classroom learning. INFACT also has materials for robotics, coding, and other everyday activities that employ computational thinking.

The scaffolds we designed to support learners in computational thinking (see Figures 12.3–12.5) are grounded in Vygotsky's notion of a ZPD where students can reach with appropriate scaffolds. We built attention highlighters, and graphical

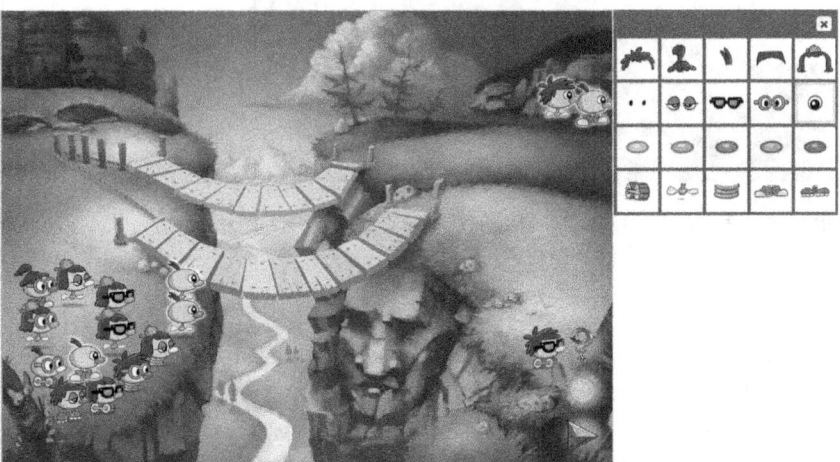

FIGURE 12.3 A highlighter tool helps players keep track of all the zoombinis with the specific value of the attribute they are examining

FIGURE 12.4 A bookkeeping tool helps players keep track of the values and attributes they already identified as causing sneezes or not in the puzzle

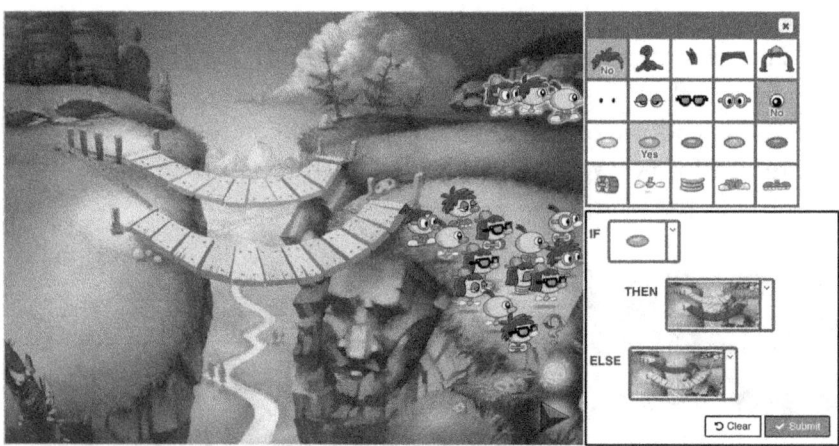

FIGURE 12.5 An expression tool supports metacognition by asking learners to articulate the rule that they discovered in the puzzle

organizers for working memory—just as Miaka suggested—as well as explicit prompts to make learners' computational thinking more visible and support metacognition.

INFACT also provides differentiation strategies to support neurodiversity. These strategies include supports for executive function and metacognition as well as multiple contexts (e.g., art, music, sports), multiple modalities (e.g., coding, robotics, physical activities), and multiple entry points into a problem. An important aspect of this differentiation work was inspired by the research of our colleague, Dr. Maya Israel, a researcher in special education in computational thinking at the University of Florida. Israel suggested providing multiple entryways into activities for different learners. For example, in an introductory coding activity, one student may start with an open-ended problem while another is provided the first few steps to get them started, and another may receive a completed program with mistakes that they need to debug. All three of these activities work toward the same goal of helping students learn to code for a given activity, but they are differentiated to cater to each learner's preparation and interests.

In an efficacy study of INFACT, students in grades 3–5 who used INFACT teaching and learning materials performed better on external measures (gains between pre and posttests) of computational thinking as compared to those in comparison classes using other computational thinking activities. We also found that students who scored in the lower half of the class on executive function tasks were the ones who grew the most in computational thinking skills using INFACT. This shows promise that executive function supports may narrow the gap for neurodivergent learners in a computational thinking school-based program. When used in concert with other supports (e.g., audio and text support) and innovative forms of assessment, we may finally be able to make headway in providing more meaningful, inclusive STEM problem-solving experiences for all. This research has built an exciting basis for our team and others to find new ways of educating exactly the innovative problem solvers we need to tackle our grandest challenges.

Bibliography for Chapter 12

Asbell-Clarke, J., Rowe, E., Almeda, V., Edwards, T., Bardar, E., Gasca, S., Baker, R. S., & Scruggs, R. (2021). The development of students' computational thinking practices in elementary-and middle-school classes using the learning game, Zoombinis. *Computers in Human Behavior, 115*, 106587.

Baron-Cohen, S. (2020). *The pattern seekers: How autism drives human invention*. Basic Books.

Barr, V., & Stephenson, C. (2011). Bringing computational thinking to K–12: What is involved and what is the role of the computer science education community? *ACM Inroads, 2*(1), 48–54.

Beck, J. E., & Gong, Y. (2013, July). Wheel-spinning: Students who fail to master a skill. In *International conference on artificial intelligence in education* (pp. 431–440). Springer International Publishing.

Begel, A., Dominic, J., Phillis, C., Beeson, T., & Rodeghero, P. (2021, March). How a remote video game coding camp improved autistic college students' self-efficacy in communication. In *Proceedings of the 52nd ACM technical symposium on computer science education* (pp. 142–148).

Brennan, K., & Resnick, M. (2012, April). New frameworks for studying and assessing the development of computational thinking. In *Proceedings of the 2012 annual meeting of the American Educational Research Association, Vancouver, Canada* (pp. 1–25).

Computer Science Teachers Association. (2017). *Computational thinking standards*. https://www.csteachers.org/page/standards

Grover, S., & Pea, R. (2013). Computational thinking in K–12: A review of the state of the field. *Educational Researcher, 42*(1), 38–43.

Israel, M., Pearson, J. N., Tapia, T., Wherfel, Q. M., & Reese, G. (2015). Supporting all learners in school-wide computational thinking: A cross-case qualitative analysis. *Computers & Education, 82*, 263–279.

Shute, V. J., Sun, C., & Asbell-Clarke, J. (2017). Demystifying computational thinking. *Educational Research Review, 22*, 142–158.

Stevenson, J. L., & Gernsbacher, M. A. (2013). Abstract spatial reasoning as an autistic strength. *PLoS One, 8*(3), e59329.

Walkowiak, E. (2021). Neurodiversity of the workforce and digital transformation: The case of inclusion of autistic workers at the workplace. *Technological Forecasting and Social Change, 168*, 120739.

Wing, J. M. (2006). Computational thinking. *Communications of the ACM, 49*(3), 33–35.

13

Neurodiversity in STEM—The Actions We Take

Reflections

The schoolchildren who are in elementary school today will be the ones responsible for solving world problems in just a few short decades. Each child comes to school as an individual—each with a particular set of cognitive strengths and weaknesses (on any particular day) and each with their own host of social, emotional, and physical circumstances. All of these individual traits must be considered and reflected in the learning process. In an ideal world, a teacher is able to differentiate learning opportunities and supports for each of these unique students—providing individualized experiences that meet each learner's particular combination of strengths, weaknesses, interests, and preparation. This is no small task, and our current educational systems are far from supporting that ideal.

As problems loom larger, and resources seem scarcer, people seem to gravitate to short-term solutions that make their lives easier now. The path of least resistance is often the status quo or a change to something more efficient and more economical. Unfortunately, in education, this just kicks the can down the road. What are little problems in young learners can become bigger problems as they mature. We need to change our systems and our culture to start thinking differently about our neurodivergent learners. We need to stop focusing on how to fix them, and think more about how to nurture their valuable contributions to an innovative problem-solving community. This is critical in light of the grand challenges we face in our future.

The stories and the research in this book are not intended to present a magic bullet or a panacea. I doubt the strategies suggested here are the only ways to approach this challenge, or that they will work for everyone, but they offer some ideas about how we might reform the way we all look at neurodiversity. I hope they spark a conversation about other methods and innovations to aid in the important next steps towards inclusion and nurturing of neurodiversity in STEM education.

Strategies for Employers

When diversifying their workforces, many STEM companies found they had to manage expectations of managers, employees, and parents of potential employee candidates when increasing their neurodivergent staff. Companies often offer workshops on communication and sensitivities for their employees, and strategies for facilitating meetings and workflows for their managers. Managers learn how to monitor their neurodiverse employees' stress, which could be due to sudden changes in work schedules or sensory overload in a busy office, and how to recognize an issue early and deal with it proactively rather than letting it brew. In other words, they become better managers.

STEM companies have found that addressing issues to support neurodivergent employees provides a wider benefit to the entire company. The need for clear communication pushes managers to put forth ideas more directly and visually, and to codify organizational processes in ways that maximizes the potential of all employees. Some companies have found that initiatives to aid in communications with autistic employees, such as increased attention to nuance, irony, and other fine points of language, have improved communication overall with all employees. Managing neurodiverse employees forces managers to shift their thinking towards optimizing conditions for each employee. This attitude results in greater productivity for each individual and the company.

Some of the strategies employers can use to help neurodiversity thrive in the STEM workplace include:

- Clearly written agendas in advance of meetings so people can gather their thoughts on topics beforehand
- Clearly written meeting notes after meetings to ensure a shared understanding of decisions and action items
- Facilitation methods to distribute speaking times so everyone has a chance to voice their opinions
- Agreed upon nonjudgmental cues for behaviors that may be unintentionally disrupting the flow of a meeting or working session
- Allowing input from employees on the types of roles they want within a team and clearly shared expectations for those roles
- Provision for autonomy of thought where employees are allowed space to work on a problem their own way before having to explain their thinking to others

Strategies for Teachers

There is plenty of evidence from research in psychology, neuroscience, and learning sciences about how students learn. It is time that we build schools, curriculum, assessments, and educational systems that are designed for the learner instead of trying to build learners to fit into our systems. We should not be asking singers to stop singing, painters to stop painting, or systematizers to stop systematizing. Each learner should be encouraged to express themselves the way they are most inclined, and then we should teach them within that space.

I know some readers are out there asking, "but don't they need to learn the other things too?" And my answer is yes. But first they need to learn how to learn. They need to feel like a learner. They need to love learning. They need to identify with learning, feel agency in their learning, and develop confidence in their ability to learn.

I had an epiphany when I first started working with Ms Bradbury's students at Hillside Junior High. I realized that if a student comes into class in a fight-or-flight state—or if they are in a fit of rage or feeling alienated out of shame and stigmatization—then nothing I do pedagogically is going to help them learn STEM. Their brain is not ready to take in new information. They are not a learner in that state. I realized I had to meet the student where *they* were, not where I decided the starting point should be. I had to let them find themselves as a successful learner and build that identity before I try to make them do new stuff that is hard for them and that they may not even care about.

Education is not solely about subject matter or the knowledge and skills of a discipline, it is also about emotion and identity and agency. It is about helping a learner see themselves as a scientist, or mathematician, or engineer, even if only for the duration of a STEM activity. They need to understand how they like to solve problems and what they are good at. We all do things differently, and the more we know about our own ways of solving problems, the better we can hone our talents and mitigate our challenges. And the more we will want to persist.

Education is also about supporting executive function and other neurological functions so that all learners can work to their potential. It is about realizing that just because a student struggles in one area, it doesn't preclude brilliance in another. Just because I need eyeglasses to see words on a page or a scratch pad to record numbers when I do calculations doesn't make me unable to do research. I just need some supports. It is our job as educators to mine for the brilliance in each student and to bring it to light. And that is a very hard job, so systems need to be much more supportive of those who teach.

Students crave a sense of belonging in the classroom. They want to understand their role and be appreciated for taking their role seriously. A move towards social–emotional learning in curriculum is focusing on fostering a sense of well-being and belonging in students. In 1997, the US Association for Supervision and Curriculum Development and the Collaborative for the Advancement of Social and Emotional Learning (CASEL) published guidelines for social–emotional learning. In it they wrote:

Schools play an essential role in preparing our children to become knowledgeable, responsible, caring adults. Knowledgeable. Responsible. Caring. Behind each word lies an educational challenge. For children to become knowledgeable, they must be ready and motivated to learn, and capable of integrating new information into their lives. For children to become responsible, they must be able to understand risks and opportunities, and be motivated to choose actions and behaviors that serve not only their own interests but those of others. For children to become caring, they must be able to see beyond themselves and appreciate the concerns of others; they must believe that to care is to be part of a community that is welcoming, nurturing, and concerned about them. Each element of this challenge can be enhanced by thoughtful, sustained, and systematic attention to children's social and emotional learning.

To create this sense of belonging in STEM problem solving, we need to connect with learners. Learning activities themselves should be relevant for students, which requires differentiation based upon their interests and lived experiences, and also should provide flexible supports and structures. Ms Bradbury used a "Getting to Know You Survey" with her incoming students. She went home and studied their responses until she had at least one interest or connection about each student that she could remember. Whether it was a shared love of cars or pizza or something exotic she wanted to learn more about, she referred to the tidbit frequently with the student. This not only gave her a foothold to connect new content to their lived experiences, but also let each student know she cared enough to learn something about them. That went a long way in cultivating the trust she earned from nearly every student.

We also need teaching and learning experiences that encourage new ideas and allow for nonconformity. We need to motivate learners and give them space for autonomy of thought. We need to provide supports for executive function and other scaffolds within the context of the STEM learning experience. And we need to assess learning using flexible and rigorous methods that look at the problem-solving process, not just product. We need to differentiate and optimize learning experiences for each individual learner. I believe that when we can engage learners in differentiated and supportive activities that foster agency and autonomy, along with STEM problem-solving experiences that foster creativity, systems thinking, and innovation, we may have a hope of nurturing the problem solvers we need for the future.

Some of the strategies educators can use to help neurodiversity thrive in the STEM classroom include:

- Provide students a choice of contexts and tools used in learning experiences
- Enable open-learning opportunities where learners can find their strengths through interest-based learning
- Use engaging contexts that immerse learners in interest-driven learning (e.g., game-based learning and project-based learning)
- Use disciplinary contexts that align with the cognitive strengths of many neurodivergent learners (e.g., computational thinking)
- Provide structures that support planning, organization, and monitoring of tasks (e.g., milestones and daily planning sheets)
- Provide scaffolds to support working memory (e.g., graphical organizers), attention (e.g., highlighting tools), and metacognition (e.g., explicit expression tools)
- Use individualized assessments that focus on process rather than product.

Strategies for Designers, Administrators, and Policy Makers

Just as the top global companies in STEM are seeing neurodiversity as a competitive advantage, our education systems must step up to the responsibility of nurturing the unique strengths of each learner. Our educational systems must recognize the vast variety of ways people learn, and they must promote teaching methods, school structures, assessment and accountability structures, and professional growth of educators with a value towards this diversity. It is not only the right thing to do, it is the right thing to do for our future economy. We need these problem solvers and we need educational systems to help them thrive, not just survive.

Inclusive Schools

Thomas Hehir and Lauren Katzman, authors of the book *Effective Inclusive Schools*, stated, as their main takeaway, that leadership matters. Leaders who value diversity in schools and set a culture within the school to do the same. The schools they studied had principles and policies within the school to recognize the importance of diversity, such as a rule that there would be no tolerance for disparaging remarks about students or families in the schools. Classes were often co-taught with both general education (non-special education) teachers and special educators working side by side. These schools often rejected a one-to-one paraprofessional model where the student was isolated with the aide, preferring integration of students within the class activities along with ample supports. This required a great deal of staff planning time, which teachers were sufficiently allotted for that purpose. They often found the school day too short to incorporate all the different supports some students needed, so they offered after-school time as well for students to receive extra help. Teachers worked hard in these schools, but they were also provided with tools, respect, and opportunities for professional growth. The leaders of the successful inclusive schools studied by Hehir and Katzman were persistent and willful. They "did not accept the default but rather established a strong inclusive direction and found the resources to make that happen." These leaders were able to work skillfully within the political frame of their educational system. And they fought hard. For example, one district felt strongly that some of their students would do better on the state exams if they could be provided with text-to-speech reading supports. They felt that their nonreaders were being unfairly assessed in other areas, such as math, because of their low reading skills. Through persistence, one principal was able to get his students permission to use a reading assistance program on the statewide assessments, which led to each student in the state having this option.

Building Community

School decisions are also made within the social context of their community. Nadine Bonda, former administrator and TERC board member, told me about a school for the gifted and talented that was started in the small, urban district where she was superintendent. Parents were very excited to have the school opening in the town. Bonda argued for housing the new school within an existing building that also housed the lowest-achieving school in the district. She went to the town council, who agreed to fund the program as long as it was not at the school she had chosen. They argued that parents wouldn't want to send their kids to *that* side of town. Bonda argued harder. She wanted the school there. She wanted to draw children from all of the city, their best and brightest, to that school. And she wanted to draw the resources she knew would be poured into the school to that neighborhood. As it turned out, the councillors were wrong. Parents were thrilled if their kids got into the school. The children had PE, art, and other learning experiences with the other children in the school building, integrating the school community as a whole. In that school, the teachers often received reports from the previous school that a child was very bright but couldn't learn because of behavior problems. Those kids were trouble at first, but when they were stimulated by all the opportunities and challenges at the new school, those children were able to thrive. Once the boredom was gone, so were the behavior problems.

In the same way, teachers create classroom communities grounded in respect "in which all students are affirmed for their value, with shared norms and responsibilities for all members". This attitude must extend to the school and the school's relationship with the community. School structures that support student learning often have collaborative teaching models with teaching teams that share students, advisory systems in which a small group of students are supported by a single advisor over multiple years, and looping, in which students stay with the same teacher for more than one year. As much as possible, the extra supports required by students are provided in class within the context of the regular lessons. Disruptions to the class time that can derail learning for many children are held at a minimum, with interruptions such as overhead announcements being saved for one part of the day.

Building community takes work. A principal who was at Hillside Junior High before I started working with Ms Bradbury had noted that the divisions between the school and community were making it very difficult to make a connection with the students. She realized that many of the students' parents had negative experiences with schooling, and authority in general, so trust was going to have to be earned before she could start building relationships. It also didn't help, she figured, that the entire teaching staff, including herself, were white when they taught so many students of color. This principal spent months going to the neighborhood churches, sometimes the only white person in the building. She sat quietly and just joined the community in their rituals. Soon someone approached her, then another started a conversation, and she started becoming a known entity in the pews of the church. Next, she wandered down into the public housing community, going to events, and bringing some clothes and food when she saw a family in need. As she continued these practices over time, she noticed more parents venturing up to the school to talk with a teacher or to pick up their child. Each time she saw a parent in the school, the principal made a point of welcoming them and showing them around.

The culture of the school and relationships between parents and the school were sustained by the teachers even after this principal left. Enough of the "old timers" saw the value of the earlier principal's investment with the community, so they maintained open communication and a welcoming attitude to the school. Teachers beamed when they saw one of their students, often a newly landed Canadian, wearing a sweater they'd handed down. Their impact on these families often extended far past the school doors.

Community groups, such as the Phoenix Youth Programs in Halifax, are another essential connection between school and community. These programs not only shelter youth who cannot live at home, or who do not have a home, but they also run tutoring and other academic programs to support the cognitive and emotional needs of their clients. By joining their tutoring program as a volunteer, I was able to support students with the same work I'd seen in class. I knew what the teacher wanted and where the student needed help. I also was able to learn more about students in their out-of-school settings, which helped my work with them in school. This home–school connection helped make students feel safe and valued.

During the onset of COVID, those first few weeks of lockdown brought many educators to the doorsteps of their students' homes, delivering food and school supplies. When children couldn't go to school, school came to the children. This was a shining example of how schools and communities can work together in the best interest of the learners and their families. When we can stitch together schools, community groups, therapists, medical programs, and all the other supports children need in their lives—when we all work together in the same direction, as a system—I believe the changes we need will come.

Some of the strategies that administrators and policy makers can use to help neurodiversity thrive in the STEM classroom, workplace, and general public include:

- Interest-driven pedagogies in schools and out-of-school programs (e.g., game-based learning and project-based learning) and providing educators with ample preparation to differentiate experiences for each learner
- Flexibility in school hours and structure to meet the needs for learner-centered pedagogies such as project-based learning
- Time and resources for co-teaching models between specialists in neurodiversity and STEM educators
- Professional growth time and resources for general education teachers to learn about executive function and neurodiversity
- Ample (i.e., way, way more) mental health and social work professionals working along with educators to support the social and emotional needs of learners in and outside the classroom.

What Can Be

When I dream ahead to what can be in our educational systems, I think back to a few wonderful examples I've seen over my career. In particular, I remember visiting an elementary school in Tampare, Finland. Children were up and moving about in their adjoining classrooms, freely moving among stations. There was an area for reading books, a game-based math program set up on a few computers, a sand and water station, and other tables where carefully designed activities immersed a wide variety of learners in rich and meaningful experiences. Children were learning collaboratively and in the ways that they chose.

I asked the teacher about special education and learning specialists, and she told me they didn't use words like that. To them it was all just "learning." They didn't distinguish children according to their learning deficits; they offered a wide set of multi-faceted learning activities with flexible supports and extensions so that students could engage at their own level of interest and ability. Children were supported for their accomplishments, not labeled by their difficulties.

I wonder how many neurodivergent learners would have an easier time in a school that was designed for them—not a place with exclusively neurodivergent learners—but a general education setting that was built with many different interests, motivations, strengths, and challenges taken into account right from the start. There are wonderful schools like that, but they are often prohibitively expensive for many families, and they also do not give everyone the benefit of inclusive learning where all talents are allowed to shine. Inclusive education should not be a modification of something else; it is a foundation that shapes everything. Everyone is front and center in an inclusive classroom.

I believe that this type of inclusive education is not only important, it is necessary. It is only by providing education that reveals, nurtures, and highlights the strengths of innovative problem solvers that we will navigate through the choppy waters ahead. We need a collaborative system, like the one Hutchins described to navigate a naval ship into harbor, where information is seen from all sides and shared cohesively among all members of the educational system. We need proactive leaders who understand where we need to go and are determined to find a path to get there. Many of these leaders are likely neurodivergent themselves.

Working alongside neurodivergent learners for the past few years, I have learned that I have so much more to learn. I am excited to discover all that our brightest, least conforming, and most creative, innovative, and rigorous thinkers will do. I am excited that the world seems to be coming to the realization that neurodiversity is an asset—a competitive advantage. And I am excited that neurodivergent learners themselves are finding a voice to make us all more aware of what they have to contribute to our society. I look forward to the day when some of the huge gnarly problems that keep me awake at night are in the hands of the brilliant neurodivergent learners I see—like Caleb and Eli and Renee and Miaka—those brilliant learners who are coming up through the ranks. I hope that our educational systems will be ready for them. And if this book can start that conversation, I will be satisfied.

So, what are you thinking?

Keep the conversation going. Ask a colleague or a friend to about how you can make change together.

Bibliography for Chapter 13

Cipriano, C., Strambler, M. J., Naples, L., Ha, C., Kirk, M. A., Wood, M. E., Sehgal, K., Zieher, A., Eveleigh, A., McCarthy, M., Funero, M., Ponnock, A., Chow, J., & Durlak, J. (2023, February 2). Stage 2 report: The state of the evidence for social and emotional learning: A contemporary meta-analysis of universal school-based SEL interventions. https://doi.org/10.31219/osf.io/mk35u

Greenberg, M. T. (2023). *Evidence for social and emotional learning in schools*. Learning Policy Institute. https://doi.org/10.54300/928.269

Hehir, T., & Katzman, L. I. (2012). *Effective inclusive schools: Designing successful schoolwide programs*. John Wiley & Sons.

Immordino-Yang, M. H., Darling-Hammond, L., & Krone, C. (2018). *The brain basis for integrated social, emotional, and academic development: How emotions and social relationships drive learning*. Aspen Institute.

For Product Safety Concerns and Information please contact our EU
representative GPSR@taylorandfrancis.com
Taylor & Francis Verlag GmbH, Kaufingerstraße 24, 80331 München, Germany

www.ingramcontent.com/pod-product-compliance
Lightning Source LLC
Chambersburg PA
CBHW050637300426
44112CB00012B/1829